Edmond Hoyle

An Epitome of Hoyle with Beaufort and Jones's Hoyle

improved

Edmond Hoyle

An Epitome of Hoyle with Beaufort and Jones's Hoyle improved

ISBN/EAN: 9783337379889

Printed in Europe, USA, Canada, Australia, Japan

Cover: Foto ©berggeist007 / pixelio.de

More available books at **www.hansebooks.com**

AN
EPITOME
OF
HOYLE,
WITH
Beaufort and Jones's Hoyle
IMPROVED;
OR,
PRACTICAL TREATISES
ON THE
FOLLOWING GAMES,

HAZARD,	DRAUGHTS,
BACKGAMMON,	WHIST,
TENNIS,	QUADRILLE,
BILLIARDS,	PIQUET,
CRICKET,	LANSQUENET, and
CHESS,	QUINZE.

With an Account of the prefent fashionable Game called E-O, played at moft of the polite Chocolate Houfes, never before attempted in Print. Comprifing the Laws and Rules of the feveral Games, as fettled at White's, Stapleton's, &c. &c. Alfo the moft advantageous Method of Betting at thofe Games, and the erroneous Odds introduced in former Productions of a fimilar kind, rectified.

BY A MEMBER OF THE JOCKEY CLUB.

DUBLIN;
PRINTED BY R. M. BUTLER, Bookseller, No. 31, GRAFTON-STREET.

M.DCC.XCI.

R E A D E R.

THE Editor of this production thinks it necessary to preface it with an Address to the
Reader, to set forth what claim it may have to the
patronage of the Public, being the first attempt of
the kind ever offered to them. Hoyle's Games are
so universally known, and have been so long in general estimation, that it is not astonishing two other
works upon the same plan, have been within these
few years published under the title of Hoyle's Games
Improved. The Editor will not pretend to detract
from their merit, but must observe, he does not
think Hoyle's Games were, in either of those productions, Improved: though it must be admitted,
that several additional Games were in each of those
books introduced, which rendered them more uni-
versal, and of course more instructive. But the
chief complaint that has ever been made against
Hoyle, is, that he is too prolix and perplexed; and
that his book is of such a size, that it cannot be in-
serted in a common pocket book. It was to obviate
this principal objection to Hoyle's Games, that the
present production was compiled, as its size will
allow it to be inclosed, without swelling to any considerable degree the common pocket book, that is

A   annually

annually published : an advantage that muft be obvious, when it is confidered that by this means, it
may be confulted as a memorandum book, when a
doubt arifes concerning the odds, or the laws of any
game, without alarming the company by producing
a volume, that muft at firft fight determine the bearer
to be, if not a pretended connoiffeur, at leaft an
ignoramus in thofe games at which he plays or bets :
a circumftance that would afford the real adepts an
opportunity of taking advantage of his ignorance.

Whilft this fmall book contains the effence of all
the games that Mr. Hoyle has written upon, it alfo
comprifes all the games that have ever been publifhed
in that form, with this difference, that the Editor
has felected from all, and given a greater diverfity
than can be met with in any other compilation. He
has alfo rectified the odds of feveral games, which
he knew to be erroneous, and fupplied them from
his own knowledge and memory.

Billiards have been treated upon in the two productions he has alluded to, but in both they are de-
fective : in the one the *lofing carambole* is not introduced ; in the other the *Ruffian carambole* is omitted.
In this little work accounts of both are to be met
with ; and the odds (erroneous in both thofe books)
are rectified.

Befides all the games treated upon by Hoyle, and
both the Hoyle's Improved, the Editor has laid before his Readers an account of the game of E-O,
the

the moft fashionable and polite now upon the *ton,* which never before was touched upon.

From what has been faid, and the great difference of the price, *three fhillings* and *one fhilling and fix-pence,* added to the advantage of this little work being fo portable and convenient, the Editor flatters himfelf, the public in general, and the admirers of the different games here treated upon, will not hefitate a moment in giving the preference to this publication, before any other of the kind that has been offered to the public.

We fhall leave it to the reader's judgment to diftinguifh the fuperiority of the ftyle between this little production and any other of the former, Hoyle's original work not excepted. The language in all thofe works are frequently obfcure, and often perplexed. The repetitions are irkfome to the eye and ear; and the number of falfe concords that prevail in two of them,* are difgufting to a perfon the leaft acquainted with grammar. Add to this their calculations are frequently erroneous, to a degree, that even a fchool boy, who had only got into addition, would be whipt for, as blunders of this kind are frequently to be met with: not to be tedious, we fhall only cite one in the game of Piquet, where fpeaking of the three hazards, namely, the pique, repique, and capot, and ftating how the cards may come to produce this effect with their value as follows:

* Hoyle and Jones.

Point

Point	—	3
Tierce major		12
Quatorze ace		14
Ditto king	—	14
Ditto queen	—	14
By play	—	13
Capot	—	40

	110

One of the improvers upon Hoyle makes the total to amount to 201, the other to 170.

The Editor cannot conclude this Addrefs without affuring his Readers, that previous to its publication, he had the honour of receiving the approbation of feveral members of Arthur's, Stapleton's, Bootle's, Brooke's, the Jockey Club, and of many other gentlemen converfant in all the games here treated upon; who have pronounced it the moft concife and complete Epitome of Hoyle and Hoyle Improved, with additions and corrections, that has ever been publifhed.

EPITOME

HOYLE's. GAMES, &c.

The Game of Hazard.

WE shall begin with the game of Hazard, as one of the most fashionable, which has been long in vogue, and formerly received the sanction of the Groom Porters, where it was publicly played; but now is confined to private parties, except at Newmarket and other races.

This game is played with a pair of dice, and not confined to any particular number of persons; from two to fifty may play at it. The player who begins, throws what is called a main, namely, a chance for the whole party: this must be above four, and not more than nine; whence it follows, he must continue his throws till he brings either five, six, seven, eight or nine; after which he must throw his own chance, which may be any above three, but must not exceed ten. If he should bring two aces, or deuce ace, usually called *crabs*, he loses his bets, let the company's chance be what it may. If the main be seven, and seven or eleven thrown immediately af-

ter,

ter, this is called a nick, and the cafter, or prefent player, wins out his ftakes. If eight be the main, and eight or twelve are thrown immediately after, it is likewife ftyled a nick, and the cafter wins his ftakes. The player throwing any other numbers for the main, fuch as are allowed, and brings the fame number the enfuing caft, it is alfo called a nick, and confequently wins all the ftakes he has made. The cafter, upon winning three fucceffive mains, pays half a guinea to the box, or proprietor of the place.

The fignification of a bet or ftake at this game, fomewhat differs from any other. If a byftander propofes laying any fum of money with the cafter, he muft depofit his cafh upon the table, within a circle which is defcribed for that purpofe, after which, if the cafter agrees to it, he knocks the dice box upon the table, at the perfon's money with whom he propofes betting, or particularly afcertains at whofe money he throws, which ratifies the wager; and he is obliged to anfwer whatever fum is down, unlefs the ftaker calls to cover, in which cafe the cafter is obliged to ftake alfo, otherwife the bets become void. It is at the option of the perfon who bets, to bar any throw which the cafter is going to throw, provided neither of the dice is vifible. If one dice fhould be feen, the cafter muft throw the other at it.

The ufual odds are as follow. If feven be thrown for a main, and four the chance, it is 2 to 1 againft the perfon that throws: if fix to four are thrown, the odds are five to three, feven to fix, 3 to 2, barring the two trois; with the two trois, only 6 to 5: feven to five, 3 to 2: fix to five, an even bet, barring the doublets or the two trois; with the trois, 5 to 4: eight to five an even bet, barring the two fours; 5 to 4 with two fours: nine to five even: nine to four, 4 to 3: the nick of feven, 7 to 2, but

oftener

oftener laid 10 to 3; and 5 to 1, that six or eight are not nicked.

The following table will be a still farther illustration of these calculations.

Table of the Odds.

7 to 4	are	2 to 1	
6 to 4	are	5 to 3	
5 to 4	are	4 to 3	
7 to 9	are	3 to 2	
7 to 6	are	3 to 2	barring two trois.
		6 to 5	with the two trois.
7 to 5	are	3 to 2	
6 to 5	are	even, barring two trois.	
		5 to 4	with the two trois.
9 to 5	are even.		
9 to 4	are	4 to 3	

The nick of seven is 7 to 2, but oftener laid 10 to 3. These calculations should be learned with attention. But this handy Pocket Companion will supply the place of any forgetfulness, as it may be so easily referred to without even the bystanders perceiving that it is called in to aid the memory, a circumstance greatly in its favour, and which could not be done by lugging out a book of a larger bulk upon similar occasions. The additional utility of knowing these odds accurately, is that of making hedges, in case the chance happens not to be a favourable one; for by taking the odds, a ready better often secures himself, and often reduces part of his wager to a certainty. Example: suppose seven to be the main, and four the chance, and the player has five pounds depending on the main, by taking six pounds to three, he must inevitably win one or two pounds; and, *vice,* if he does not approve his chance, by laying the odds against himself, he must secure part of the bet, in proportion to what he originally laid.

The

The Game of Backgammon.

AS Backgammon is a game played with a pair of dice, and the only one of the kind we shall introduce, we have placed it next to Hazard. It is played by two persons only, on a table divided into two parts, containing 24 black and white points. Each player has 15 men, the one black, the other white, by way of distinction, and they are disposed of in the following manner. If you play into the right hand table, two men are placed upon the ace point in your adversary's table, five upon the sixth point in the opposite table, three upon the cinq point in the hithermost table, and five upon the sixth point in your own table. The chief object is to bring the men round in your own table ; consequently all throws that tend to this pursuit, and at the same time impede your adversary in executing the same design, are in your favour ; the contrary success of your opponent must of course be against you. The first most advantageous throw is aces, as it blocks the sixth point in the outer table, and secures the cinq point in your own ; and your adversary's two men upon your ace point cannot escape, with his throwing either quatre, cinq, or fix. This established advantageous throw is, therefore, often asked and given by way of odds, from a superior to an inferior player.

As it is necessary for a learner to know how many points he ought to throw upon the two dice upon an average, we shall take the following method of illustration.

EXAMPLE. I would know how many chances there are upon the two dice? Answer, 36. I would also know how many points there are upon 36 chances?

Answer,

Answer.

2 Aces	—	4	5 and 4 twice	18
2 Deuces	—	8	5 and 3 twice	16
2 Trois	—	12	5 and 2 twice	14
2 Fours	—	16	5 and 1 twice	12
2 Fives	—	20	4 and 3 twice	14
2 Sixes	—	24	4 and 2 twice	12
6 and 5 twice	22		4 and 1 twice	10
6 and 4 twice	20		3 and 2 twice	10
6 and 3 twice	18		3 and 1 twice	8
6 and 2 twice	16		2 and 1 twice	6
6 and 1 twice	14			

Divided by 36 { 294 / 288 } 8 Points.

6.

294 divided by 36, solves the question, whereby it is proved, that one throw with another, you may expect to throw 8 upon two dice.

Would you know how many chances there are upon two dice? The answer is 36, which are as follow:

2 Sixes	—	1	5 and 4 twice	2
2 Fives	—	1	5 and 3 twice	2
2 Fours	—	1	5 and 2 twice	2
2 Trois	—	1	5 and 1 twice	2
2 Deuces	—	1	4 and 3 twice	2
2 Aces	—	1	4 and 2 twice	2
6 and 5 twice	2		4 and 1 twice	2
6 and 4 twice	2		3 and 2 twice	2
6 and 3 twice	2		3 and 1 twice	2
6 and 2 twice	2		2 and 1 twice	2
6 and 1 twice	2			

36

The

The foregoing, which are the principal calculations at Backgammon, will convey a sufficient idea of the chances of dice to a beginner, until by practice, he becomes a proficient sufficient to enter into the more abstruse supputations: we shall, therefore, now enter upon giving a notion of the game more at large. If you play three up at Backgammon, your chief design should be in the first instance, either to secure your own, or your adversary's cinq point, which being effected, you may play a pushing game, and strive to gammon your opponent.

After having gained your cinq point, the next advantage is to gain your bar point, as it will prevent your adversary's moving with sixes.

Having proceeded thus far, you are to prefer making your quarter point in your own tables, rather than the quarter point out of them.

These points being attained, you have a very probable chance of gammoning your opponent, if he be very forward. For example: suppose his tables are broken at home, it will then be for your advantage to open your bar point, and compel him to come out of your tables with a six, and having your men dispersed, you may not only take up that man which your adversary brings out of your tables, but you will likewise have a good chance of taking the man left in your tables, supposing your adversary had two men there, after you had made up your bar point. If he should have a blot in his own tables, it will not be your interest to make up your tables; because if he should enter upon a blot, which you should intentionally make, you may probably get a third man, in which case, you will have at least 4 to 1 of the gammon; when, on the other hand, if you have only two of his men up, the odds are in his favour that you do not gammon him.

H

If you play for a single hit, the taking of only one or two men, renders it more secure than a greater number, supposing that your tables are well made.

Directions for carrying your men home.

In carrying your men home, that you may not lose a point, you should carry your most distant man to your bar point, as the first stage to place it on. The next stage is six points farther, namely, in the first place when your adversary's are first placed out of his tables. The third stage is the sixth point in your own tables. This method should be pursued till all your men are brought home, except two, when by leaving a point, you may often save your gammon, by the assistance of two fives or two fours.

When you play only for a single hit, your adversary's cinq point is the chief object to obtain. If you should not succeed in this respect, by being hit by your adversary, and you find he is more forward than yourself, you should throw more men into his tables, in the following manner : place a man upon your cinq or bar point, and if your adversary should not hit it, you may then gain a forward, instead of a back game : but if he should take you up, you must play for a back game ; and in that case, the greater number of men which are taken up, render your game the better, as you will thereby be enabled to preserve your game at home : and you should then always endeavour to gain both your adversary's ace and trois points, or his ace and deuce points, and carefully preserve three men upon his ace point, that in case you should hit him from thence, that point may still remain secure.

Never play for a back game at the beginning of a set, as this would be very disadvantageous, by running the risk of being gammoned in pursuit of a single hit.

B 2 *Directions*

Directions for playing at starting the 36 *chances.*

Two aces are to be played on your cinq point and bar point for a gammon, or for a hit.

Two aces to be played on your adverſary's bar point, and on your own bar point, for a gammon, or for a hit.

Two trois to be played on your cinq point, and the other two on your trois point in your own tables, for a gammon only.

Two fours to be brought over from the five men placed in your adverſary's tables, and to be put upon the cinq point in your own tables, for a gammon only.

Two fives to be brought over from the five men placed in your adverſary's tables, and to be placed on the trois point in your own tables, for a gammon, or for a hit.

Six ace, take your bar point for a gammon or for a hit.

Six deuce, a man to be brought from the five men placed in your adverſary's tables, and to be placed on the cinq point in your own tables, for a gammon or for a hit.

Six and trois, a man to be brought from your adverſary's ace point, as far as it will go for a gammon or for a hit.

Six quatre, a man to be brought from your adverary's ace point, as far as it can go for a man or for a hit.

Six cinq, a man to be carried from your adverſary's ace point, as far as it will go for a gammon or a hit.

Cinq and quatre, a man to be carried from your adverſary's ace point, as far as he can go for a gammon or for a hit.

Cinq trois, make the trois point in your table, for a gammon or for a hit.

Cinq

Cinq deuce, play two men from the five placed in your adverfary's table, for a gammon or a hit.

Cinq ace, bring one man, from the five placed in your adverfary's tables for the cinq, and play one man on the cinq point in your own tables for the ace, for a gammon only.

Quatre trois, bring two men from the five placed in your adverfary's tables, for a gammon or for a hit.

Quatre deuce, make the quatre point in your own tables, for a gammon or for a hit.

Quatre ace, play a man from the five placed in your adverfary's tables for the quatre, and for the ace play a man upon the cinq point in your own tables, for a gammon only.

Trois deuce, play two men from the five placed in your adverfary's table, for a gammon only.

Trois ace, make the cinq point in your own tables, either for a gammon or for a hit.

Deuce ace, play one man from the five men placed in your adverfary's tables for the deuce ; and for the ace, play a man upon the cinq point in your own tables, for a gammon only.

The laws of Backgammon.

1ft. If you take a man from any point, that man muft be played.

2d. You are not underftood to have played any man, till you have placed him upon a point, and quitted him.

3d. If you play with 14 men only, there is no penalty attending this miftake, becaufe by playing with a lefs number than the complement, you play to a difadvantage, not having the additional man to make up your tables.

4th. If you bear any number of men, before you have entered a man taken up, and which you were neceffarily compelled to enter, fuch men, fo borne,

muft

muſt be entered again in your adverſary's tables, as well as the man taken up.

5th. If you ſhould miſtake your throw and play it, and your adverſary afterwards throws before he diſcovers the error, the throw cannot be recalled with-out the conſent of both parties.

The Game of Tennis.

THE uſual extent of a Tennis Court is about 96 feet in length, and 33 in breadth. It is divid-ed into equal parts by a net, over which the ball muſt be ſtruck, to have any advantageous effect to the players. When you enter a court, there is a long gallery covered with a penthouſe, that muſt be paſſed before the *dedans* is reached. This latter place the ſpectators and betters uſually remain in, whilſt a match is playing. The paſſage is divided into different compartments, which are ſtyled galleries. From the line towards the *dedans* is the firſt gallery, the door ſecond gallery, and the laſt gallery, which is called the ſervice ſide. From the *dedans* to the laſt gallery, the figures 1, 2, 3, 4, 5, 6, are inſcribed at a yard diſtance each, which mark the chances, that conſti-tute one of the moſt eſſential parts of the game. On the other ſide of the line is the firſt gallery; door, ſe-cond gallery, and laſt gallery. This is called the ha-zard ſide. Every ball played into the laſt gallery on this ſide tells for a certain ſtroke, the ſame as into the *dedans*. Between the ſecond and the laſt gallery are the figures 1, 2, marking the chances on the ha-zard ſide. Over this long gallery, as was previouſly mentioned, is the penthouſe, upon which the ſervice is given to begin the ſet; and if the player miſſes

ſtriking

striking the ball so as to rebound from the penthouse over a certain line, it is styled a *fault*, two of which amount to the loss of a stroke. If the ball passes round the penthouse, on the opposite side of the court, and falls beyond a particular described line it is called *passe*, reckons for nothing, and the player must again serve.

Upon the right side of the court from the *dedans*, a part of the wall projects more than the rest, in order to variegate the strokes, and make them more difficult to be returned, which is called the *tambour*, alluding to the effect of a drum. The grill is a kind of window in the right-hand corner of the hazard side: when a ball is lodged here, it is reckoned a determinate stroke, without depending upon chaces.

The usual set of Tennis consists of six games; but when an advantage set is played, two successive games above five must be won to decide it; or when it becomes six games all, two additional games must be won, without intermission, to determine the set.

Upon the player's giving the first service, his adversary must return the ball, wherever it falls after the first rebound. Example; if at the figure 1, the chace is called at a yard, that is to say a yard from the *dedans*, the chace remains till a second service is given; and if the player on the service side, lets the ball go after his adversary returns it, and if the ball falls on or between any of these figures, the players change sides; as the first player will then be on the hazard side to play for the first chace, which if he wins, by striking the ball so as to fall, after its first rebound, nearer to the *dedans* than the figure 1, without his adversary's being able to return it from its first rebound, he wins a stroke, and proceeds in like manner to win a second stroke, &c. If a ball falls on a line with the first gallery, door, second gallery, or last

gallery,

gallery, the chace is called at such a place, naming the gallery, &c.   When it is just put over the line, it is called a chace at the line.   If the player on the service side returns a ball with such force as to strike the wall on the hazard side, so as to rebound, after the first hop over the line, it is also called a chace at the line.

The hazard side chaces arise from the ball being returned either too hard, or not hard enough, so that the ball, after its first rebound, falls on this side of the line which describes the hazard side chaces, in which case it is a chace at 1, 2, &c. provided there be no chace depending, and according to the spot where it actually falls.   When the opponents change sides, the player, in order to win his chace, must put the ball over the line, any where, so that his adversary does not return it.   When there is no chace on the hazard side, all balls reckon that are put over the line from the service side without being returned.

Tennis, instead of being marked in the usual numerical gradations, is called for the first stroke 15, for the second 30, for the third 40, and for the fourth game; unless the players make four strokes each; then instead of calling it 40 *all*, it is called *deuce*; after which, the next stroke is called *advantage*; and in case the strokes become once more equal, *deuce* again, until one or the other of the players obtains two strokes successively to complete the game.

The precise odds at Tennis are not ascertained with that precision as might be wished for, on account of the chances, and the variety of methods of giving odds to render a party nearly equal.

The lowest odds given consist in a *bisque*, (except the choice of the sides of the court), which amounts to the liberty of scoring a stroke whenever the player, who receives this advantage, pleases to demand it.

Example ;

Example; suppose a game to be forty or thirty, he who is forty, and entitled to the *bisque*, gets the game.

General rules of odds besides.

Fifteen is a stroke given at the commencement of a game.

Half thirty, is fifteen given the first game, and *thirty* the second game, and thus progressively to the whole thirty, forty, &c. &c.

Half court, implies the player being compelled to play into the adversary's half court.

Touch no wall, is a still greater advantage, tho' the former is very considerable.

Round service, is giving a service all round the extremity of the penthouse.

Barring the hazards, is foregoing to reckon the dedans tambour, or last gallery on the hazard side, &c.

The usual odds that are laid, allowing for a variety of circumstances, are as follow:

After the first stroke being won, which is called 15, the odds become

Of the single game —	7 to 4
Thirty love — —	4 to 1
Forty love — —	8 to 1
Thirty fifteen —	5 to 1
Forty thirty —	3 to 1

The odds of a four game set, when the first game

is won, are — —	7 to 4
When two games love	4 to 1
Three games love —	8 to 1
When two games to one	2 to 1
Three games to one —	5 to 1

The odds of a six game set, when the first game

is won, are — —	3 to 2
When two games love	2 to 1
Three games love —	4 to 1

C   Four

Four games love —	10 to 1
Five games love —	21 to 1
When two games to one	8 to 5
Three games to one —	5 to 2
Four games to one —	5 to 1
Five games to one —	15 to 1
When three games to two	7 to 4
Four games to two —.	4 to 1
Five games to two —	10 to 1
When four games to three	2 to 1
Five games to three —	5 to 1

The odds of an advantage set, when the first game

is won, are — —	5 to 4
When two games love —	7 to 4
Three games love —	3 to 1
Four games love —	5 to 1
When two games to one	4 to 3
Three games to one —	2 to 1
Five games to one —	10 to 1
Four games to two —	3 to 2
When four games to three	8 to 3
Five games to three —	3 to 1
Two games to four —	2 to 1
Six games to five —	5 to 2

The

The Game of Cricket.

Laws of the Game, as settled by the Gentlemen of the Committee of Kent, Hampshire, Surry, Sussex, Middlesex, and London, at the Star and Garter, Pallmall.

THE ball must weigh not less than five ounces and a half. nor more than five ounces and three quarters.

It cannot be changed during the game but with consent of both parties.

The bat must not exceed four inches and a quarter in the widest part.

The stumps must be twenty-two inches, the bail six inches long.

The bowling crease must be parallel with the stumps, three feet in length, with a return crease.

The popping crease must be three feet ten inches from the wickets; and the wickets must be opposite to each other, at the distance of 22 yards.

The bowler must deliver the ball with one foot behind the bowling crease, and within the return crease; and shall bowl four balls before he changes wickets, which he shall do but once in the same innings.

He may order the player at his wicket to stand on which side he pleases.

The striker is out if the bail is bowled off, or the stump bowled out of the ground:

Or if the ball, from a stroke over or under his bat, or upon his hands (but not his wrists), his held before it touches the ground, though it be hugged to the body of the catcher:

Or if in striking, both his feet are over the popping crease, and his wicket is put down, except his bat is grounded within it:

Or

Or if he runs out of his ground to hinder a catch :

Or, if a ball is ftruck up, and he wilfully ftrikes it again :

Or, if in running a notch, the wicket is ftruck down by a throw, or with the ball in hand, before his foot, hand, or bat is grounded over the popping creafe ; but if the bail is off, a ftump muft be ftruck out of the ground by the ball :

Or if the ftriker touches or takes up the ball before it has lain ftill, unlefs at the requeft of the oppofite party :

Or if the ftriker puts his leg before the wicket, with a defign to ftop the ball, and actually prevents the ball from hitting his wicket by it.

If the players have croffed each other, he that runs for the wicket that is put down is out : if they are not croffed, he that has left the wicket that is put down is out.

When the ball has been in the bowler's or wicket keeper's hands, the ftrikers need not keep within their ground till the *umpire* has called *play* ; but if the player goes out of his ground with an intent to run, before the ball is delivered, the bowler may put him out.

When the ball is ftruck up in the running ground between the wickets, it is lawful for the ftrikers to prevent its being caught ; but they muft neither ftrike at nor touch the ball with their hands.

If the ball is ftruck up, the ftriker may guard his wicket either with his bat or his body.

In fingle wicket matches, if the ftriker moves out of his ground to ftrike at the ball, he fhall be allowed no notch for fuch a ftroke.

The wicket keeper fhall ftand at a reafonable diftance behind the wicket, and fhall not move till the ball is out of the bowler's hand, and fhall not, by any noife, intercept the ftriker : and if his hands,
                                                                    knees,

knees, feet, or head, be ever before the wicket, though the ball hit it, he shall not be out.

The umpires shall allow two minutes for each man to come in, and fifteen minutes between each innings, when the umpires shall call *play*; the party refusing to play shall lose the match.

They are the sole judges of fair and unfair play, and all disputes shall be determined by them.

When a striker is hurt, they are to allow another to come in ; and the person hurt, shall have his hands in any part of that innings.

They are not to order a player out, unless appealed to by the adversaries.

But if the bowler's foot is not behind the bowling crease, and within the return crease, when he delivers the ball, the umpire, unasked, must call *no ball*.

Bets.

If the notches of one player are laid against those of another, the bet depends upon both innings, unless specified differently.

If one party beats the other in one innings, the notches of the first innings shall decide the wager.

But if the other party goes in a second time, then the bet must be determined by the numbers on the score.

The

The Game of Billiards.

THIS game is played on a table covered with fine green cloth, about twelve feet long, and six wide, forming an exact oblong : it is surrounded with cushions to keep the balls within the table, and to cause their rebounding. There are six holes or pockets, which are to receive the balls, and when they enter the pockets they are called hazards, each of which, at the usual game, reckons two in favour of the player who puts in his adversary's ball, and on the contrary he loses two, if he puts in his own ball. Billiards are played with a mace or a cue : the first is composed of a stick about a yard and a half in length, with a head at the end : a cue is a thick stick at one end, running tapering towards the other, till it comes to a point somewhat less than a six-pence. The cue is played over the left hand, and supported with the fore finger and the thumb. Mace playing, and what was called long play, or trailing, with sticks longer than usual, was formerly in vogue ; but now this manner of playing is entirely exploded in all public and polite companies ; and the cue is the only fashionable instrument used, being by far the most agreeable, fairest, and ingenious game, requiring much more address and attention than the mace, played either long or short.

General Rules observed at the common Game of Billiards.

For the lead, the balls must be placed at one end, and the player must strike them against the farthermost cushion, in order to see which will rebound nearest the cushion that is next to them.

The nearest to the cushion is to lead, and choose which ball he pleases. The plain ball is generally chosen, as sometimes the spot on the marked ball becomes an index for a hazard.

The

The leader is to place his ball at the stringing nail, and not to pass the middle hole by following the ball with his mace or butt end of his cue : and if he loses himself in leading, he loses the lead, which is an advantage to a judicious player.

The next player must stand within the corner of the table, and not place his ball beyond the nail.

He who plays upon the running ball loses one, as does he who touches the ball twice ; but these last severities are seldom played.

He who does not hit his adversary's ball loses one. He who touches both balls at the same time makes a foul stroke, in which case, if he should hole his adversary, nothing is gained by the stroke ; but if he should put himself in, he loses two.

He who holes both balls loses two.

He who strikes upon his adversary's ball, and holes himself, loses two.

He who plays at the ball without striking it, and holes himself, loses three.

He who strikes his own, or both balls over the table, loses two; and if his own ball goes over the table without touching his adversary's, he loses three.

He who retains the end of his adversary's stick when playing, or endeavours to balk his stroke, loses one.

He who takes up his own ball, or his adversary's, without permission, loses one.

He who plays another's ball or stroke, without leave, loses one.

He who stops either ball when running, loses one ; and being near the hole, loses two.

He who shakes the table when the ball is running, loses one.

He who strikes the table with the stick, or plays before his turn, loses one.

He

He who throws the stick upon the tables, and hits either ball, loses one.

If the ball stands upon the edge of the hole, and after being challenged it falls in, it reckons nothing, but must be placed where it was before.

If any person, not being one of the players, stops a ball, it must stand on the place where it was stopt.

He who plays without a foot upon the floor, and holes his adversary's ball, gets nothing for it, but loses the lead.

He who leaves the game before ended, loses it.

Any person may change his mace or cue, and is allowed long sticks and butts, when the ball is out of reach.

If any dispute arises between the players, the marker, or the majority of the company, who are not betters, must decide it.

Those who are not players must stand from the table, and make room for the players.

If any person lays any wager, and does not play, he shall not give advice respecting the game, on penalty of paying all the bets depending.

The common winning game is played with two white balls, the one having only a small black speck to distinguish it from the other; so is the losing game, the winning and losing, choice of balls, bricole, the bur hole, and the four game; but hazards are played with as many balls as players, who never exceed six; and the carambole, now much in vogue, is played with three balls, one of which is red; and this game is also diversified, as we shall hereafter mention.

Fortification Billiards, which were formerly played at a table near St. James's-Square, and which we believe to have been the only one that ever existed, have long been exploded as puerile, and only fit to amuse children.

The *losing game* is the usual game reversed; for except hitting the ball, which is essential, the player
gains

gains every advantage by losing. When he holes himself he gains two; if he puts in his adversary's ball alone, he loses two; but when both balls go in, the player marks four. This game in a great measure depends upon strength, and a knowledge of the bricole, or the rebounding of the player's ball from the cushion, which constitutes some of the finest strokes that are played at it. The losing game is necessary to be known to play with judgment the winning game, in guarding against the danger of losing one's self.

The *winning and losing game* is constituted by blending the two games together, as all balls that are holed after striking the adversary's ball, reckon to the advantage of the player; consequently holing both balls is scored four. When the balls go over the cushions, either at this or the losing game, no advantage arises from it.

Choice of balls, is taking each time which ball the player chooses, which being so capital an advantage, is usually played against winning and losing.

Bricole is playing the striker's own ball against the cushion previously to its hitting the adversary's ball: if it does not touch upon the rebound, the player loses one. When played against the common game, it is esteemed so great a disadvantage, that eight or nine points are usually given the bricole player.

The *bar hole* is thus intitled, from the pocket being barred for which the adversary's ball should be played, and the striker being obliged to play for another hole. This game requires great judgment of the doubles and round abouts from the cushions; and the knowledge of it is of great advantage to the player of the common game, as there are many balls that should not be played for the first hole that presents itself, as being dangerous, either from being what

D   is

is called a spread eagle or a pair of breeches; fine cuts for the middle hole when peril stares one in the face from the corner, or a dead full ball that is likely to be followed. The disparity between the bar hole and the common white game, between equal players, is calculated at about six or seven the most.

The *one-hole* game is to the ignorant an entire deception. As all balls which go into the one-hole reckon, the player of that game aims to lay his ball constantly before that hole, and his antagonist is often embarrassed to keep both balls out of that hole, particularly upon the leads, when the one hole player constantly endeavours to place his ball before it, if not on the brink of the hole.

The *four game* consists of two partners opposed to two others at the usual white winning game, who play successively after each hazard, or the loss of two points. This game is played fifteen points up, whence it arises that the point or hazard becomes an odd number, consequently a miss is of more importance at this game than any other.

Hazards derive their name from their depending entirely upon hazards being made, without any regular game interfering. As has already been mentioned, seldom more than six persons play. A fixed sum is named for each hazard, and the miss is half. Every player whose ball is holed pays that sum; and if he misses the ball he proposes playing upon, he pays the moiety of the loss of the hazard. Seldom much money is played for at hazard, they being considered as a mere pastime till a regular match can be made. However, some general rule is usually observed at this desultory game; which is never to lay a hazard, if it can be avoided, for the next player; and this may in a great measure be obviated by playing upon his ball, and either placing him close to the cushion, or at a

consider-

confiderable diftance from thofe balls that are in danger of being next holed. As no regular game is played, the table is paid for by the hour.

· We now come to the three different kinds of *carambole* game, namely, the ordinary carambole the
Ruffian carambole, and the carambole lofing game.
In thefe games there is fuch a variety of chances, and
indeed what may be ftyled luck, that they are fo very
uncertain, as to preclude the calculation of any regular
odds, which when laid, are nothing more than the
effects of caprice, or the ufual cuftom of the table. ·

Carambole is a new-fangled game of French ex-
traction, as its name implies. It is played with three
balls, two white and one red; the latter being neu-
tral, and never played with, and is, at ftarting, placed
on a fpot marked in the center between the two
ftringing nails, at the farther end of the table from
whence the players begin their game, when their
balls are ftruck from a mark parallel to the carambole.
This is alfo obferved at the beginning of every hazard.
The principal object of this game is for the player to
ftrike with his own ball the two others, which ftroke
is called a carambole, and marks two points, part of
fixteen which conftitute the game. If the player holes
the red ball he fcores three, and upon holing his adverfary's ball he gains two; and thus it frequently
happens that feven are got upon a fingle ftroke, by
caramboling and holing both balls. There is a great
analogy between this game and the lofing, and it is
ufually played with the cue.

The fecond object at this game, after making what
is called the *carambole*, is the *baulk*. This confifts of
the player's making the white ball, and bringing his
own ball and the carambole within the ftringing
nail, where the opponents firft began. In confe-
quence of this, the laft player's adverfary is compel-

led

led to play bricole from the lower cushion : if he
misses both the white and red ball, he loses one, and
probably leaves the red ball an immissible hazard,
from which circumstances the game is often immediately determined.

The *Russian carambole*, which is seldom played, is
an intended refinement or improvement upon the
former game. At this game the red ball is placed on
the mark as at the simple carambole ; but the player
at the beginning of the game, or after having been
pocketed, never plays from any particular spot, but
is at liberty to place the ball where he chooses.
When the game begins, the first player does not
strike at the red ball, but places his own as nearly
as he can behind the carambole, and then his adversary is at liberty to play at which he pleases : if he
plays at the red ball and pockets it, he reckons three,
as in the former game, towards twenty points, of
which the Russian carambole consists ; when the red
ball is replaced upon the spot on which it was first
fixed, when he may strike again, or take his choice
which of the two balls to play at, always pursuing
his stroke till both balls are made. The player gains
two by caramboling, and loses as many as he might
have got, if he caramboles and holes himself : for instance, if he, the player strikes, at the red ball and
holes it, and at the same time caramboles and holes
himself, he loses five ; and if he holes both balls when
he carambules, and likewise his own, he loses seven,
which he would have gained if he had not lost himself. It varies very little in other respects from the
original carambole.

Carambole losing game. This begins in the same
manner as the carambole winning game, and sixteen
is the number. If the striker misses both balls he
loses a point. If he misses both balls and holes it, he
loses

lofes three. If he hits the red ball firft and holes it, he lofes three. If he holes the red and white balls by the fame ftroke, he lofes five. If he makes a carambole, and holes either the adverfary's or the red ball only, he gets nothing for the carambole, and lofes either two or three, according to which ball he played upon. A carambole reckons two. If the ftriker makes a carambole by hitting the white ball firft, and fhould lofe himfelf, he gets four. If he makes a carambole by ftriking the red ball firft, and holes himfelf, he gets five. If he makes a carambole by ftriking the white ball firft, and holes himfelf and his adverfary, he gets fix. If he makes a carambole by ftriking the red ball firft, and holes himfelf and his adverfary, he gets feven. If he makes a carambole by ftriking the white ball firft, and holes himfelf and the red ball, he wins feven. If he makes a carambole by ftriking the red ball firft, and holes himfelf and the red ball, he wins eight. The reader will eafily fupply all the other ftrokes that can occur at this game, by comparing it with thofe that precede it.

Odds at Billiards, the white-game, equal players.

1 love is 5 to 4		2 to 1 is 4 to 3
2 love is 3 to 2		3 to 1 is 3 to 2
3 love is 7 to 4		4 to 1 is 7 to 4
4 love is 2 to 4		5 to 1 is 2 to 1
5 love is 5 to 2		6 to 1 is 7 to 2
6 love is 4 to 1		7 to 1 is 4 to 1
7 love is 9 to 2		8 to 1 is 9 to 1
8 love is 10 to 1		9 to 1 is 10 to 1
9 love is 15 to 1		10 to 1 is 50 to 1
10 love is 60 to 1		11 to 1 is 60 to 1
11 love is 63 to 1		

3 to 2 is	5 to 4		6 to 5 is	3 to 2
4 to 2 is	8 to 5		7 to 5 is	7 to 4
5 to 2 is	7 to 4		8 to 5 is	3 to 1
6 to 2 is	5 to 2		9 to 5 is	4 to 1
7 to 2 is	7 to 2		10 to 5 is	5 to 1
8 to 2 is	6 to 1		11 to 5 is	21 to 2
9 to 2 is	7 to 1			
10 to 2 is	21 to 1		7 to 6 is	4 to 3
11 to 2 is	23 to 1		8 to 6 is	2 to 1
			9 to 6 is	5 to 2
4 to 3 is	4 to 3		10 to 6 is	5 to 1
5 to 3 is	8 to 5		11 to 6 is	6 to 1
6 to 3 is	5 to 2			
7 to 3 is	3 to 1		8 to 7 is	6 to 4
8 to 3 is	6 to 1		9 to 7 is	2 to 1
9 to 3 is	7 to 1		10 to 7 is	9 to 2
10 to 3 is	20 to 1		11 to 7 is	5 to 1
11 to 3 is	21 to 1			
			9 to 8 is	4 to 3
5 to 4 is	5 to 4		10 to 8 is	11 to 4
6 to 4 is	7 to 4		11 to 8 is	3 to 1
7 to 4 is	2 to 1			
8 to 4 is	4 to 1		10 to 9 is	9 to 4
9 to 4 is	9 to 2		11 to 9 is	5 to 2
10 to 4 is	21 to 2			
11 to 4 is	12 to 1		11 to 10 is	5 to 4

This table of the general odds will give the reader
a competent idea of all the compound odds, which
are all formed upon the same principle, but are sel‑
dom laid.

The Game of Chess.

THE game of Chefs is played upon a common French draught board, containing fixty-four fquares. The king and his officers, confifting of eight pieces, are placed upon the firft line of the board, the white corner being placed towards the right hand. The white king is placed upon the fourth black fquare, the black king upon the fourth white fquare, oppofite each other. The white queen upon the fourth white fquare, on the left of her king. The black queen upon the fourth black fquare, on the right of her king. The bifhops on each fide of the king and queen. The knights each fide of the bifhops. The rooks in the two corners of the board, next to the knights. The eight pawns to be placed upon the eight fquares of the fecond line. The pieces and pawns on the fide of the king, derive their names from him, as thofe on the fide of the queen do from her, and are called the pawns of the bifhop, of the king, or of the bifhop of the queen. The fquares borrow their names from the pieces, viz. that in which the king ftands is called the king's fquare; where the pawn ftands is called the king's fecond fquare; that beyond it is named the king's fourth fquare, and fo on. The king moves every way, but only one fquare at a time. He may leap once in the game, either on his own fide, or on the queen's, (viz. the rook is moved into the next fquare on the other fide of him, which is alfo called cuftling), provided no piece is between him and the rook; nor after this rook has been moved; nor after the king has been moved; nor when the king is in check; nor when the fquare over which he means to leap is viewed by an adverfe man, who would check him in his paffage. The kings muft conftantly

constantly be one square distant from each other. The
queen has the move of the rook and the bishop, moving in a streight line, and also diagonally. The bishops
move only angularly. The knights move in an
oblique manner upon every third square, from
black to white, and *vice versa*, over the heads of the
men. The rooks move in a direct line. A pawn
moves in a strait line forward, and takes the enemy
angularly. He can move two squares the first move.
If the square over which a pawn leaps, is viewed by
an opposite man, that man may take the pawn in his
passage, and then he must place himself upon the
square over which the pawn leaped. A pawn can
move only one square at a time, after the first move.
The rest of the men move forward or backward.
When a pawn attains the head of the board upon the
first line of the adversary, he may be changed for any
one of the pieces, which have been lost. The men
take the adversary's men who stand in their way,
when the road is open for them, or they reject it, if
the player judges proper. A man should be set down
in the same square in which the contrary man is
taken. The men can move the whole length or
breadth of the board, or from one angle to the other,
except the kings, knights and pawns. When the
adversary king is in a situation to be taken, the player
must say *check* to him, by which he receives warning
to defend himself, either by moving, or covering himself by one of his own men, or taking the man who
assaults him : if he cannot effect either of these ob-
jects, he is *checkmated*. The king cannot change his
square, if by this operation he goes into a check.
When the king has no man to play, and is not in
check, yet is blocked up in such a manner that he cannot move without going into check, this situation is
stiled a *stalemate* ; and in this case the king who is
thus *stalemated* wins the game.

Extracts

Extracts of Rules and Observations for Chess, by the late
Mr. H O Y L E.

You ought to move your pawns before you stir your pieces, and afterwards to bring out to support them : therefore the kings, queens, and bishops pawns, should be first played, in order to open your game well.

You are not therefore to play out any of your pieces early in the game, because you thereby lose moves, in case your adversary has it in his power, by playing a pawn upon them to make them retire, and also opens his game at the same time : especially avoid playing your queen out, till your game is tolerably well opened.

Avoid giving useless checks, and never give any, unless you thereby gain some advantage, because you may lose the move, if he can either take or drive your piece away.

Never crowd your game by having too many pieces together, for fear of choaking up your passage, so as to hinder your advancing or retreating your men as occasion may require.

If your game happens to be crowded, endeavour to free it by making exchanges of pieces or pawns ; or castle your king as soon as you can conveniently.

Endeavour to crowd the adversary's game, which is to be done thus : when he plays out his pieces before he does his pawns, you are to attack them as soon as you can with your pawns, by which you may make him lose moves, and consequently crowd him.

Never attack the adversary's king without a sufficient force ; and if he attacks your king, and you have it not in your power to attack his, you are to offer exchanges with him ; and if he retires when you present a piece to exchange with him, he may

E lose

lose a man, and consequently you will gain an advantage.

Play your men in such good guard of one another, that if any man you advance be taken, the adverse piece may also be taken by that which guarded yours; and for this purpose, be sure to have as many guards to your pieces as you see your adversary advances pieces upon it, and if you can, let them be of less value than those that he assails with.

If you find that you cannot well support your piece, see if by attacking one of his that is better, or as good, whether you cannot thereby save your own.

Never make an attack but when well provided for it, nor give useless checks, for thereby you open the adversary's game, and make him ready prepared to pour in a strong attack upon you, as soon as your weak one is over.

Never play any man till you have examined whether you are free from danger by your adversary's last move; nor offer to attack, till you have considered what harm he would be able to do you by his next moves, in consequence of your own, that you may prevent his designs if dangerous.

When your attack is in a prosperous way, never be diverted from pursuing your plan, if possible, to give him mate, by taking any piece, or other advantage your adversary may purposely throw in your way, with the intent that by snapping at that bait, he might gain a move that would frustrate your design.

In pursuing a well laid attack, when you find it necessary to force your way through your adversary's defence, with the loss of some pieces, if, upon counting as many moves forward as you can, you perceive a prospect of success, rush on, boldly, and sacrifice a piece or two to gain your end. It is from these bold attempts that the finest games are produced.

Never

Never let your queen ſtand before your king, as that your adverſary, by bringing a rook or biſhop, might check your king if ſhe were not there, for you might ſcarcely have the chance to ſave her.

Let not your adverſary's knights, eſpecially if duly guarded, come to check your king and queen, or your two rooks at the ſame time; as in the two firſt caſes, the king being forced to go out of check, the queen or the rook muſt be loſt; and in the two laſt caſes, a rook muſt at leaſt be loſt for a worſe piece.

Take care that no guarded piece belonging to your adverſary forks two of your pieces.

When the kings have caſtled on different ſides of the board, the adverſary muſt advance upon the other king the pawns he has on that ſide of the board, taking care to bring his pieces, eſpecially the queen and rooks, to ſupport them; and the king who has caſtled, is not to ſtir his three pawns till forced to it.

In playing the game, endeavour to have a move as it were in ambuſh; that is to ſay, place the queen's biſhop or rook behind a pawn, or piece, in ſuch a manner, as that upon playing that pawn or piece, you diſcover a check upon your adverſary's king, and conſequently may often get a piece, or ſome other advantage by it.

Never guard an inferior piece with a better, if you can do it with a pawn, becauſe the better piece may in that caſe be, as it were, out of play: for the ſame reaſon you ought not to guard a pawn with a piece, if you have it in your power to guard it with a pawn.

A pawn paſſed and well ſupported, often coſts the adverſary a piece. And if you play to win the game only, whenever you have gained a pawn, or

any

any other advantage, and are not in danger of lofing the move by it, make as frequent exchanges of pieces as you can.

If you have three pawns each upon the board, and no piece, and you have one of your pawns on one fide of the board, and the other two on the other fide, and your adverfary's three pawns are oppofite to your two pawns, march with your king, as foon as poffible, to take his pawns; and if he goes with the king to fupport them, go on to the queen with your fingle pawn, and if he attempts to prevent him, take his pawns, and pufh the others to the queen.

At the latter end of the game, each party having only three or four pawns on different fides of the board, the kings fhould endeavour to gain the more, in order to win the game. EXAMPLE: if you bring your king oppofite your adverfary's king, with only one houfe between you, you will have gained the move.

When your adverfary has his king and one pawn on the board, and you have your king only, you will never lofe that game, if you can bring your king to be oppofite to your adverfary's, when he is either immediately before, or on one fide of his pawn, and there is only one houfe between the kings.

When your adverfary has a bifhop and one pawn on the rooks line, and his bifhop is not of the colour that commands the corner houfe his pawn is going to, and you have only your king, if you can get into that corner you cannot lofe the game, but may win it by a ftale-mate.

If you have greatly the difadvantage of the game, having only your queen left in play, and your king happens to be in the pofition of ftale-mate, continue giving check to your adverfary's king, taking great care not to check him where he can introduce any

of

of his pieces that conſtitute ſtale-mate. By purſu-
ing this method, you will at laſt compel him to take
your queen, and then you muſt neceſſarily win the
game by being in a ſtale-mate.

Never cover a check with a piece that a pawn
puſhed upon it may take, for fear of only getting
that pawn for it.

Be careful that your adverſary's king has always a
move, leſt he ſhould get a ſtale-mate; you ſhould
accordingly avoid crowding him up with pawns, for
fear you ſhould give a ſtale-mate.

Laws of Chess.

Firſt. If a player touches his man he muſt play it,
and if he quits it, he cannot recall it.

Second. If inadvertently, or otherwiſe, a falſe
man is played, and the adverſary, takes no notice
of it, till after he has played his next move, it cannot be recalled by either of the parties.

Third. In caſe a player miſplaces the men, and
he plays two moves, it is at the option of the adverſary, to puniſh him, to begin the game again or not.

Fourth. If the adverſary plays, and if even a
check, to a player's king, and gives no notice of
it, the player may let him ſtand ſtill till he gives notice of it.

Fifth. After the king is moved, the king cannot
caſtle.——The game of Cheſs requires greater attention, perhaps, than any other that is played;
and a good cheſs player will not let his thoughts be
diverted by any foreign object. To this unremitting
application, and the ſlowneſs of the moves, may be
aſcribed the frequent great length of the games.

The Game of Draughts.

General Rules for playing the Game.

The table muſt be placed with an upper white corner towards the player's right hand.

After this let the white ſquares be numbered from 1 to 32, as in the ſubjoined plan

The black men are placed upon the firſt twelve ſquares in all the ſubſequent games.

The letters N, C, F, T, at the top of each game, imply number, colour, from, to.

For the playing any number, the numbers may be inſcribed on the board itſelf near a corner of each ſquare, ſo as to be eaſily ſeen when the men are placed. Again, a table may be drawn upon paper, and the ſquares numbered as in the annexed plan.

GAME

GAME I.

N	C	F	T	N	C	F	T	N	C	F	T
1	B	11	15	19	B	11	25	37	B	14	18
2	W	22	18	20	W	32	27	38	W	16	11
3	B	15	22	21	B	5	14	39	B	7	16
4	W	25	18	22	W	27	23	40	W	20	11
5	B	8	11	23	B	6	10	41	B	18	23
6	W	29	25	24	W	16	12	42	W	11	8
7	B	4	8	25	B	8	11	43	B	23	27
8	W	25	22	26	W	28	24	44	W	8	4
9	B	12	16	27	B	25	29	45	B	27	31
10	W	24	20	28	W	30	25	46	W	4	8
11	B	10	15	29	B	29	22	47	B	31	27
12*	W	27	24	30	W	26	17	48	W	24	20
13	B	16	20	31	B	11	15	49	B	27	23
14	W	23	16	32	W	20	16	50	W	8	11
15	B	15	19	33	B	15	19	51	B	23	18
16	W	24	15	34	W	24	20	52	W	11	8
17	B	9	14	35	B	18	27	53	B	18	15
18	W	18	9	36	W	31	24	&c.	W		loses.

GAME II.

N	C	F	T	N	C	F	T	N	C	F	T
1	B	11	15	19	B	5	14	37	B	10	17
2	W	22	18	20	W	24	19	38	W	21	14
3	B	15	22	21	B	15	24	39	B	30	25
4	W	25	18	22	W	28	19	40	W	14	9
5	B	8	11	23	B	14	17	41	B	9	15
6	W	29	25	24	W	32	27	42	W	9	6
7	B	4	8	25	B	10	14	43	B	2	9
8	W	25	22	26	W	27	24	44	W	13	6
9	B	12	16	27	B	3	7	45	B	15	18
10	W	24	20	28	W	30	25	46	W	6	2
11	B	10	15	29	B	6	9	47	B	7	10
12	W	27	24	30	W	13	6	48	W	2	6
13	B	1	10	31	B	22	13	49	B	10	14
14	W	24	19	32	W	14	18	50	W	6	9
15	B	14	23	33	B	23	14	51	B	25	21
16	W	25	22	34	W	23	14	52	W	3	21
17	B	17	13	35	B	16	30	53	B	14	17
18	W	18	9	36	W	25	21	&c.			drawn.

* 12 White loses the game by this move.

GAME III.

N.	C.	F.	T.	N.	C.	F.	T.	N.	C.	F.	T.
1	B	11	15	19	B	16	23	37	B	12	19
2	W	22	18	20	W	26	10	38	W	24	8
3	B	15	22	21	B	14	23	39	B	3	12
4	W	25	18	22	W	27	18	40	W	13	9
5	B	8	11	23	B	6	15	41	B	14	18
6	W	29	25	24	W	13	6	42	W	28	24
7	B	4	8	25	B	1	10	43	B	18	23
8	W	25	22	26	W	31	26	44	W	24	19
9	B	10	15	27	B	5	9	45	B	23	27
10	W	24	20	28	W	26	23	46	W	19	15
11	B	12	16	29	B	9	13	47	B	27	32
12	W	21	17	30	W	23	19	48	W	15	11
13	B	7	10	31	B	22	13	49	B	32	27
14	W	17	13	32	W	15	22	50	W	9	5
15	B	8	12	33	B	32	28	51	B	27	25
16	W	28	24	34	W	10	14	52	W	5	1
17	B	10	14	35	B	10	19	53	B	22	26
18	W	23	19	36	W	19	16	&c.	W		drawn

GAME IV.

N.	C.	F.	T.	N.	C.	F.	T.	N.	C.	F.	T.
1	W	22	18	19	W	25	22	37	W	32	23
2	B	11	15	20	B	16	20	38	B	6	10
3	W	18	11	21	W	19	16	39	W	13	9
4	B	8	15	22	B	20	27	40	B	2	7
5	W	21	17	23	W	31	24	41	W	17	13
6	B	4	8	24	B	12	15	42	B	9	14
7	B	17	13	25	W	23	19	&c.			drawn
8	B	8	11	26	B	10	14				
9	W	25	22	27	W	17	10				
10	B	9	14	28	B	7	14				
11	W	29	25	29	W	24	19				
12	B	5	9	30	B	15	28				
13	W	23	19	31	W	28	24				
14	B	14	17	32	B	1	5				
15	W	27	23	33	W	22	17				
16	B	17	21	34	B	14	18				
17	W	22	17	35	W	26	23				
18	B	11	16	36	B	18	27				

GAME V.

N.	C.	F.	T.	N.	C.	F.	T.	N.	C.	F.	T.
1	W	22	18	19	W	25	22	37	W	9	5
2	B	11	15	20	B	7	11	38	B	19	24
3	W	18	11	21	W	24	20	39	W	5	1
4	B	8	15	22	B	15	24	40	B	11	16
5	W	21	17	23	W	28	19	41	W	20	11
6	B	4	8	24	B	10	14	42	B	7	1
7	W	17	13	25	B	17	10	43	B	16	7
8	B	8	11	26	B	24	13	44	B	16	20
9	W	25	22	27	W	13	6	45	W	5	5
10	B	9	13	28	B	1	10	46	B	24	27
11	W	29	25	29	W	22	17			drawn	
12	B	5	9	30	B	24	28	&c.			
13	W	23	19	31	W	17	13				
14	B	14	17	32	B.	3	7				
15	W	27	23	33	W	13	9				
16	B.	17	22	34	B.	16	11				
17	W	22	17	35	W.	23	16				
18	B.	11	16	36	B	12	19				

GAME VI.

N.	C.	F.	T.	N.	C.	F.	T.	N.	C.	F.	T.
1	W	22	18	19	W	24	19	37	W	27	23
2	B	11	15	20	B	15	24	38	B	16	20
3	W	18	11	21	W	28	19	39	W	31	27
4	B	8	15	22	B	6	10	40	B	6	9
5	W	25	22	23	W	22	17	41	W	18	15
6	B	4	8	24	B	13	22	42	B	9	18
7	W	29	25	25	B	26	17	43	W	23	14
8	B	8	11	26	B	11	15	44	B	12	16
9	W	23	18	27	W	32	28	45	W	19	12
10	B	9	13	28	B	15	24	46	B	10	19
11	W	18	14	29	W	28	19	47	W	12	8
12	B	10	17	30	B	1	6	&c.		drawn	
13	W	21	14	31	W	30	20				
14	B	6	10	32	B	21	8				
15	W	25	21	33	B	26	23				
16	B	10	17	34	W	8	11				
17	W	21	24	35	W	23	18				
18	B	2	6	36	B	11	16				

GAME VIII.

N.	C.	F.	T.	N.	C.	F.	T.
1	W	22	18	19	W	27	18
2	B	11	15	20	B	7	10
3	W	18	15	21	W	24	20
4	B	8	15	22	B	16	15
5	W	21	17	23	W	18	15
6	B	4	8	24	B	19	23
7	W	23	19	25	W	15	11
8	B	8	11	26	B	10	14
9	W	17	13	27	W	11	8
10	B	9	14	28	B	22	26
11	W	25	21	29	W	31	22
12	B	14	18	30	B	14	17
13	W	26	22	31	W	21	14
14	B	18	22	32	B	6	9
15	W	23	18	33	W	13	6
16	B	11	16	34	B	13	26
17	W	18	15	35	W	8	4
18	B	16	20	&c.			drawn.

GAME VII.

N.	C.	F.	T.	N.	C.	F.	T.	N.	C.	F.	T.
1	W	22	18	19	W	23	18	37	W	29	22
2	B	11	15	20	B	11	16	38	B	14	18
3	W	18	11	21	W	27	23	39	W	23	14
4	B	8	15	22	B	15	20	40	B	6	10
5	W	21	17	23	W	32	27	41	W	15	6
6	B	4	8	24	B	10	14	42	B	2	25
7	W	17	13	25	W	17	10	43	W	19	15
8	B	8	11	26	B	7	14	44	B	25	30
9	W	23	19	27	W	18	9	45	B	27	23
10	B	9	14	28	B	5	14	46	W	20	27
11	W	25	21	29	W	13	9	47	W	31	24
12	B	14	18	30	B	6	13	48	B	30	26
13	W	26	23	31	W	19	15	49	W	23	18
14	B	18	22	32	B	1	6	50	B	26	22
15	W	30	26	33	W	24	19	51	W	18	14
16	B	15	18	34	B	3	7	52	B	12	16
17	W	26	17	35	W	28	24	53	W	15	11
18	B	18	22	36	B	32	25	&c.			drawn.

G A M E IX.

N.	C.	F.	T.	N.	C.	F.	T.
1	W	22	17	19	W	30	26
2	B	11	15	20	B	6	9
3	W	25	22	21	W	19	15
4	B	8	11	22	B	11	16
5	W	29	25	23	W	25	21
6	B	9	13	24	B	16	19 *
7	W	17	14	25	W	23	16
8	B	10	17	26	B	12	19
9	W	21	14	27	W	32	28
10	B	4	8	28	B	1	6
11	W	24	19	29	W	15	11
12	B	15	24	30	B	7	16
13	W	28	19	31	W	14	10
14	B	11	16	32	B	6	15
15	W	22	18	33	W	18	11
16	B	16	20	34	B	2	6
17	W	26	22	35	W	22	16
18	B	8	11	36	B	lofes.	

* 24 Lofes the game by this move.

G A M E X.

N.	C.	F.	T.	N.	C.	F.	T.
1	B	11	15	19	B	15	24
2	W	22	17	20	W	28	19
3	B	8	11	21	B	7	11
4	W	25	22	22	W	22	18
5	B	9	13	23	B	13	22
6	W	23	18	24	W	18	9
7	B	6	9	25	B	6	13
8	W	27	23	26	W	25	18
9	B	9	14	27	B	3	8
10	W	18	9	28	W	18	14
11	B	5	14	29	B	10	17
12	W	30	25	30	W	21	16
13	B	1	6	31	B	11	14
14	W	24	19	32	W	14	9
15	B	15	24	33	B	2	7
16	W	28	19	34	W	9	6
17	B	11	15	35	B	7	10
18	W	32	28 &c.				drawn.	

The Game of Whist.

WHIST being confidered as the moft univerfal game played in polite companies, we fhall be more particular upon this head than any other, and have placed it the foremoft amongft the games at cards.

It may not be improper to premife, that the name of this game implies SILENCE : WHIST is an interjection derived from the Italian word *zitto*, which fignifies to *command filence.* The application is obvious: this game requires great attention, and uncommon application, to play it with fkill and judgment, and to advert properly to which cards have been played, and recollect thofe that remain in hand; to this end, Mr. Hoyle, and fome others, have invented what they call a Technical Memory (which we fhall introduce,) in order to fupply the place of common recollection. Moreover, the penalties of this game are very rigid upon byeftanders as well as players, who by words, or even figns, give any of the party the leaft intimation how to play ; and an uninterefted fpectator has been known many times to pay all the bets depending upon the game playing, to a confiderable amount, for having reminded one of the partners of fome card played or occurrence, that he had forgot, and which proved to the player's advantage

Whift is played by four perfons, who cut in for partners, according to the two higheft and two loweft cards, which unite them. The partners being placed oppofite to each other, the player who cut the loweft card deals firft, giving one card at a time all round, till the pack is exhaufted, when he turns up the laft card to determine the trump, which is to remain upon the table till every one has played a

card,

card, when the dealer takes it up. The left hand
adverfary to the dealer begins, and he who wins the
trick plays again. The honours are the ace, king,
queen, and knave of trumps When three of them
fall into the hands of one perfon, or two partners,
they fcore for two points; and four honours in the
fame predicament reckon four towards the game,
which confifts of ten points. The honours are not
reckoned till the tricks are counted: each trick
above fix fcores one.

General Rules for playing.

Upon firft leading, play the beft fuit you have:
having a fequence of king, queen, and knave, or
queen, knave, and ten, they are certain leads, and
never fail gaining the *ten ace*,* for yourfelf or partner in other fuits. Firft play the higheft of the
fequences, unlefs you have five, when you fhould
play the loweft, except in trumps, when the higheft
muft be played, in order to take the ace or king,
whereby you make room for your fuit,

When you have five of the fmalleft trumps, and
no good cards, in the other fuits trump about,
which will certainly have this good effect, make your
partner the laft player, and thereby give him the ten
ace.

Having only two fmall trumps, with ace and king
of two other fuits, and a deficiency in the fourth
fuit, make as many tricks as you can with all expe-
dition. If your partner rejects either of your fuits
do not force him, as that may weaken his game.

It is not neceffary to return your partner's lead
immediately, if you are in poffeffion of good fuits,
unlefs it be at a critical part of the game. By good

* Ten ace is when an intervening card is wanting to make
the fequence; example, ace, queen, knave, &c.

suits is underflood king, queen, and knave, or queen, knave, and ten.

If both parties have five tricks, and you are pretty certain of getting two tricks in your hand, fail not winning, though in expectation of fcoring two, becaufe if you lofe the odd trick, it makes two difference, and you play two to one againft yourfelf. There is, however, an exception to this rule, which is when you perceive a probability either of faving your lurch, or winning the game, in either cafe the odd trick fhould be rifked.

When there is a plaufible appearance of winning the game, always rifk a trick or two, becaufe the fhare of the ftake which the adverfary has by a new deal, is more than equivalent to the point or two you rifk in that deal.

When your adverfary is fix or feven love, and you are to lead, you fhould rifk a trick or two, in hopes of bringing the game to an equality; wherefore, if you have queen or knave, and one other trump, and no other good cards, play out your queen or knave of trumps, as you will thereby ftrengthen your partner's hand, if he is ftrong in trumps; and if he fhould be weak, you will not hurt him.

When you are four, you fhould play for an odd trick, becaufe it faves your lurch; and to this end, though you are pretty ftrong in trumps, play your trumps cautioufly. By ftrong in trumps is underftood, if you fhould have one honour and three trumps.

When you are nine, though very ftrong in trumps, if you perceive your partner has a chance of trumping any of your adverfary's fuits, do not trump about, but let him have an opportunity of trumping thofe fuits. Being one, two, or three, you fhould play quite differently; and likewife at five, fix, or feven; becaufe in thefe cafes you play for more than a fingle point.

Being

Being the laft player, and finding that the third hand cannot play a good card to his partner's lead, fuppofing you have not a good hand of your own, return the lead upon the adverfary, which will give your partner the ten ace in that fuit, and often oblige your adverfary to change fuits, and thereby gain the ten ace alfo in that fuit.

Having ace, king, and four fmall trumps, begin with a fmall one, as it is an equal wager that your partner has a better trump than the laft player : in this cafe you will have three rounds of trumps; if not, you cannot bring out all the trumps.

When you have ace, king, knave, and three fmall trumps, you fhould begin with the king, and then play the ace, unlefs trumps are refufed by one of the adverfaries, as the odds are in your favour that the queen falls.

When you have king, queen, and four fmall trumps, you fhould begin with a fmall one, as the odds are on your fide that your partner has one honour.

Having king, queen, ten, and three fmall trumps, begin with the king, as you have a good chance that the knave will fall in the fecond round ; or you may ftay to finefle your ten upon the return of trumps from your partner.

When you have queen, knave, and four fmall trumps, begin with a little one, as the odds are in your favour, that your partner is poffeffed of at leaft one honour.

If you have queen, knave, nine, and three fmall trumps, begin with the queen, as it is a fair chance that the ten will fall in the fecond round ; or you may wait to finefle the nine.

When you have knave and ten, with four fmall trumps, begin with a fmall one.

When

When you have knave, ten, eight, and three small trumps, begin with the knave, to prevent the nine from making a trick ; and there are odds in your favour that the three honours fall in two rounds.

Having six smaller trumps, you should begin with the lowest, unless you have ten, nine, and eight, and an honour is turned up against you ; in this case, if you are to play through the honour, begin with the ten, which will compel the adversary to play his honour to his disadvantage, or leave your partner to choose whether he will take it or not.

Having ace, king, and three small trumps, begin with the small one.

With ace, king, knave and two small trumps, begin with the king, which will probably inform your partner, that you have the ace and knave remaining ; and by throwing the lead into your partner's hand, he will play you a trump, after which you are to finesse the knave, and no ill consequences can arise, unless the queen lies singly behind you.

Having king, queen, and three small trumps, begin with a small one.

With king, queen, ten, and two small trumps, begin with the king.

Having queen, knave, and three small trumps, begin with a small one.

With queen, knave, nine, and two small trumps, begin with a small one.

If you have knave, ten, and three small trumps, begin with a small one.

When you have knave, ten, eight, and one small trump, begin with the knave ; because in two rounds of trumps, the odds are that the nine falls ; or, upon the return of the trumps from your partner, you may finesse the eight.

When you have five smaller trumps, you should begin with the lowest, unless you have a sequence of

ten,

ten, nine, and eight, when you fhould begin with the higheft of the fequence.

Having ace, king, and two fmall trumps, begin with a fmall one.

When you have ace, king, knave, and one fmall trump, begin with the king.

With king, queen, and two fmall trumps, begin with a fmall one.

With king, queen, ten, and one fmall trump, begin with the king, and wait for the return of trumps from your partner, when you fhould fineffe your ten, in order to win the knave.

When you have queen, knave, nine, and one fmall trump, begin with the queen, in order to prevent the ten from making a trick.

With knave, ten, and two fmall trumps, begin with a fmall one.

If you have knave, ten, eight, and one fmall trump, begin with the knave, in order to prevent the nine making.

When you have ten, nine, eight, and one fmall trump, begin with the ten, which leaves it at your difcretion to pafs it or not.

When you have ten, and three fmall trumps, begin with a fmall one.

Particular Rules to be obferved.

When you have ace, king, and four fmall trumps, with a good fuit, you fhould play three rounds of trumps, to prevent your ftrong fuit being trumped.

Having king, queen, and four fmall trumps, with a good fuit, trump out with the king, becaufe when you get the lead again, you will have three rounds of trumps.

When you have king, queen, ten, and three fmall trumps, with a good fuit, trump firft with the king,

G with

with the view of making the knave fall at the fecond round; and do not wait [to finesse the ten,
left your strong suit should be trumped.

Having queen, knave, nine, and two small trumps,
and a good suit, trump first with a small one.

With the queen, knave, nine, and two small
trumps, and a good suit, trump first with the queen,
with the view of making the ten fall at the second
round: do not wait to finesse the nine, but trump a
second time.

Having knave, ten, and three small trumps, with
a good suit, trump out with a small one.

With knave, ten, eight, and two small trumps,
with a good suit, trump out with the knave, in hopes
of the nine's falling at the second round.

With ten, nine, eight, and two small trumps, with
a good suit, begin to trump with the ten.

*Rules for playing particular Games, after a Learner
has attained to some Degree of proficiency at the Game.*

If you are elder hand, and your game consists of
king, queen, and knave, of one suit; ace, king,
queen, and two small cards, of another sort; king
and queen of a third suit, and three small trumps;
question, how is this hand to be played? Answer.
You should begin with the ace of your best suit, or a
trump, to acquaint your partner you have the command of that suit; but you should not proceed with
the king of the same suit, but play a trump next. If
you find your partner is not strong enough to support
you in trumps, and that your adversary plays to your
weak suit, viz. the king and queen only, in that case
play the king of the best suit; and if you judge
there is a probability of either of your adversaries
being likely to trump that suit, proceed to play the
king of the suit of which you have king, queen, and
knave.

knave. If your adverſaries do not play to your
weakeſt ſuit, though apparently your partner cannot
aſſiſt you in trumps, continue trumping as often as
the lead comes into your hand. By this method, if
your partner has but two trumps, and your adverſaries have four each, by three rounds of trumps there
remain only two trumps againſt you.

If you be elder hand, and have ace, king, queen,
and one ſmall trump, with a ſequence from the king
of five in another ſuit, with four other cards of no
value, begin with the queen of trumps, and purſue
the lead with the ace, which will intimate to your
partner that you have the king; and as it would be
bad play to continue trumps the third round, till you
have firſt gained the command of your great ſuit,
by ſtopping in this manner, it alſo acquaints your
partner that you have the king and one trump only
remaining; for if you had ace, king, queen, and
two trumps more, and trumps went round twice,
you could not be hurt by playing the king the third
round. When you lead ſequence begin with the
loweſt, becauſe if your partner has the ace he will
play it, which makes way for your ſuit. Having acquainted your partner with the ſtate of your game,
upon his getting the lead, if he has any trumps remaining, he will play trumps, with the probable expectation that your king will purge the adverſaries
hands of all their trumps.

If you be ſecond player, and have ace, king, and
two ſmall trumps, with a quint major of another ſuit,
in the third ſuit you have the three ſmall cards, and
in the fourth one. Your right hand adverſary begins with playing the ace of your weak ſuit, and
then plays the king. Do not upon this occaſion
trump it, but throw away a loſing card; and if he
next plays the queen, throw away another loſing
card; and continue the ſame the fourth time, with

G 3 the

the view that your partner may trump it; who will
then play a trump, or to your ftrong fuit.

Should trumps be played, continue them two
rounds, and then enter upon your ftrong fuit. By
this means, if there fhould happen to be four trumps
in one of your adverfaries hands, and two in the other;
which is probably the cafe, your partner having a
right to three trumps out of the nine, your adverfaries have only fix between them : your ftrong fuit
will force their beft trumps, and you may make the
odd trick in your hand only. On the contrary, if
you had trumped one of your adverfaries beft cards,
you would have weakened your own hand in fuch a
manner, as probably not to make more than two
tricks without your partner's aid.

If you have ace, queen, with three fmall trumps,
ace, queen, ten, and nine of another fuit, with two
fmall cards of each of the other fuits, and your partner leads to your ace, queen, ten and nine, and as
you fhould at this game rather endeavour to deceive
your adverfaries than to acquaint your partner, put
up the nine, which will naturally induce your adverfary to play trumps, if he fhould win it. When
trumps are played to you, return them upon your
adverfary; preferving the command in your own hand.
Should your adverfary who led trumps, put up a
trump, which your partner cannot win, and he has
no good fuit of his own to play, he will return your
partner's lead, thinking that fuit lies between his
partner and yours. Should this fineffe fucceed, it
muft be very advantageous to you, and can fcarce
poffibly be detrimental.

If you have ace, king, and three fmall trumps,
with a quart from a king, and two fmall cards of
another fuit, and one fmall card to each of the other
fuits, when your adverfary leads a fuit of which your
partner

partner has a quart major, your partner puts up the
knave, and plays the ace : you refuse playing to that
suit, by playing your loose card : when your partner
plays the king, your right hand adversary trumps it
with the knave or ten ; do not trump over him, as
that might probably lose you two or three tricks,
by weakening your hand ; but in case he should lead
the suit that you have none of, trump it, and play
the lowest of your sequences, in order to get the ace
out of your partner or adversary's hand ; then as
soon as you have the lead, play two rounds of trumps,
and continue playing your strong suit. If your adversary, instead of playing to your weak suit, should
play trumps, continue the two rounds, and next endeavour to get the command of your strong suit.

*Games to be played when your Adversary on your right
Hand turns up an Honour, with Instructions how to
play when an Honour is turned up on your left Hand.*

If the knave is turned up on your right hand, and
you have king, queen, and ten, with the view of
winning the knave play the king first, which will induce your partner to think you have the queen and
ten remaining, particularly if you have a second
lead, and you do not proceed to play the queen.

In the same situation, having the ace, queen, and
ten, by playing your queen, it appears to the same
purpose as the former rule.

When the queen is turned up on your right hand,
and you have ace, king, and knave, by playing the
king, the purpose of the former rule is answered.

When an honour is turned up on your left hand,
and you hold no honour, you should play trumps
through that honour ; but in case you hold an honour, the ace excepted, you must be cautious how
you

you play trumps, for if your partner has no honour,
your adverfary will retort your own game upon you.

Cafes for demonftrating the Danger of forcing your Partner.

A, B, are partners. A has a quint major in
trumps, with a quint major and three fmall cards of
another fuit. A has the lead; and the adverfaries,
C and D, have only five trumps in either hand; in
this cafe, A having the lead, wins every trick.

On the contrary, if C has five fmall trumps, with
a quint major, and three fmall cards of another fuit,
and C has the lead, who forces A to trump firft, by
this means A wins only five tricks.

The Advantage of a Saw.

A and B are partners: A has a quart major in
clubs, being trumps, another quart major in hearts,
another quart major in diamonds, and the ace of
fpades. The adverfaries, C and D, have the following cards, namely, C has four trumps, eight hearts
and one fpade; D has five trumps, and eight diamonds; C leads a heart, D trumps it; D plays a
diamond, C trumps it; and thus continuing the faw,
each partner trumps a quart major of A's, and C being to play at the ninth trick, plays a fpade, which
D trumps: thus C and D win the nine firft tricks,
and leave A with his quart major in trumps only.

The foregoing cafe evidently proves the advantage
of purfuing a faw when once eftablifhed.

Directions for playing when an Ace, King, or Queen, is perceived on your right Hand.

If the ace is turned up on your right hand, and
you have the ten and nine of trumps only, with ace,
king, and queen, of another fuit, and eight other
cards

cards of no value, queſtion, how ſhould this game
be played ? Anſ. Begin with the ace, which inti-
mates to your partner, that you are in poſſeſſion of
the command of that ſuit. Afterwards play your
ten of trumps, as it is five to two that your partner
has king, queen, or knave of trumps; and though
it is near ſeven to two he has not two honours, yet if
he ſhould chance to have them, and they are the
king and knave, your partner will paſs your ten of
trumps; and as it is thirteen to twelve againſt the
laſt player's holding the queen of trumps, if your
partner has it not, when your partner has the lead,
he will play to your ſtrong ſuit; and upon your having the lead, you ſhould play the nine of trumps,
which enables your partner to be pretty near certain
of winning the queen, if he lies behind her.

Should the king or queen be turned up on the right
hand, the like method of play may be purſued; but
you ſhould always diſcriminate with reſpect to your
partner's abilities, as a good player will make a proper uſe of ſuch play; but it will ſeldom, if ever, be
of any ſervice to a bad one.

If the adverſary on your right hand leads the king
of trumps, and you ſhould have the ace and four
ſmall trumps, with a good ſuit, in this ſituation it is
your intereſt to paſs the king; and though he ſhould
have the queen and knave of trumps, with one more,
if he be a tolerable player, he will play the ſmall one,
upon the ſuppoſition that his partner may have the
ace. If he plays the ſmall one, you are to paſs it,
as it is an equal chance that your partner has a better trump than the laſt player: if this ſhould be the
caſe, and he is not a bad player, he will judge you
have grounds for playing in this manner, and therefore if he has a third trump remaining he will play
it, otherwiſe his beſt ſuit.

A critical

A critical Cafe to win an odd Trick.

A and B are partners againft C and D, and the
game is nine all, with the trumps all played out; A
being the laft player has the ace, and four other
fmall cards of a fuit, in his own hand, and a thirteenth card remaining; B has only two fmall cards
of A's fuit; D has king, knave, and one fmall card
of the fame fuit. A and B have won three tricks;
C and D have won four tricks; hence it follows, that
A muft win four tricks out of the fix remaining
cards to avoid lofing the game; C leads this fuit,
and D plays the king; A gives him that trick, D
returns the fame, A paffes it, and C plays his queen.
C and D have now won fix tricks, and C fancying
his partner has the ace of that fuit returns it, whereby A wins the four laft tricks, which make him
game.

If you fhould hold the king and five fmall trumps,
and your right hand opponent plays the queen, do
not in this cafe play your king, as it is an equal
chance your partner has the ace; and if your adverfary fhould have king, knave, ten, and one fmall
trump, it is likewife an equal bet that the ace lies
fingle, either in your adverfary or partner's hand;
in either cafe, it would be bad play to put on your
king: but if the queen of trumps fhould be led,
and you have the king, with two or three trumps, it
is judgment to put on the king, as it is good play
to lead from the queen, and one fmall trump only.
In this cafe, if your partner has the knave of trumps,
and your left hand adverfary holds the ace, you lofe
the trick by not putting on the king.

The ten or nine being turned up on your right hand, &c.

If the ten is turned up on your right hand, and
you have the king, knave, nine, and two fmall
trumps, with eight other cards of no value, and you
fhould

should lead trumps, begin with the knave, in order
to prevent the ten from making; and though it is
only about five to four that your partner holds an
honour, yet if that should fail, by finessing your
nine, on the return of trumps from your partner,
the ten remains at your devotion.

When the nine is turned up on your right hand,
and you have knave, ten, and eight, and two small
trumps, by leading the knave, it answers the same
purpose as in the preceding case.

There is a great difference between a lead of
choice, and a forced lead of your partner; because
in the first instance, it is supposed he leads from his
best suit, and discovering your deficiency in that suit,
and not being sufficiently powerful in trumps, nor
willing to force you, he will play his next best suit;
by which change of play, it is almost demonstrable
that he is weak in trumps. Should he persist in
continuing his first lead, supposing him a good
player, you are to conclude he is strong in trumps,
and it is a clue for you to play your game accordingly.

It is particularly detrimental at this game, to
change suits frequently: because in every fresh suit,
you risk giving your adversary the ten ace; consequently, though you lead, and are possessed of the
queen, ten, and three small ones of that lead, and
your partner plays the nine only, if you happen to
be weak in trumps, and you have no tolerable suit
to lead from, your best play is to pursue the lead of
that suit by playing your queen, which leaves it in
your partner's choice, whether he will trump it or
not, in case he should have no more of that suit; but
upon your second lead, if you should have queen or
knave of any other suit, and one card only of the
same suit, it would be judicious to lead from your
queen or knave of either of these two suits, the odds

H being

bein five to two that your partner has one honour at least in either of those suits.

When you have ace, king, and a small card, of any suit, with four trumphs, should your right hand adverfary lead that fuit, pafs it, as it is an equal wager your partner has a better card in that fuit than the third hand; in this cafe you gain a trick; if otherwife, having four trumps, you need not be under any apprehenfions of lofing by it, as when the trumps are played, it may be fuppofed the long trump is in your hand.

A Cafe which often happens.

If you have two trumps remaining, when the adverfaries have only one, and your partner appears to have a ftrong fuit, you fhould play trumps, although you may have the worft, in order to pave the way for your partner's fuit, by extracting the trumps from your adverfaries.

The Method of playing the Sequences.

Play the higheft in fequences of trumps, unlefs you have ace, king, and queen; in that cafe play the loweft, which acquaints your partner with the ftate of your hand.

If you have the king, queen, and knave, and two fmall ones, which are not trumps, begin with the knave, whether ftrong in trumps or not, as he will make way for the whole fuit to bring the ace out.

Being ftrong in trumps, and having a fequence of queen, knave, and ten, with two fmall cards of a fuit, you fhould play the higheft of your fequence; for if either of your adverfaries fhould trump that fuit in the fecond round, you being alfo ftrong in trumps, will make the remainder of that fuit by fetching out their trumps. You may play in the
like

like manner when poffeffed of the knave, ten, and nine, and two fmall cards of the fame fuit.

Having king, queen, and knave, with one fmall card of any fuit, whether ftrong in trumps or not, play the king; and when there are only four in number, the fame method fhould be obferved with inferior fequences.

If you are weak in trumps, begin by the loweft of the fequence, becaufe if your partner fhould have the loweft of that fuit, he will make it. Should you have the ace and four fmall cards of a fuit, and be weak in trumps, leading from that fuit you fhould play the ace. When ftrong in trumps, the game may be played in a different manner.

<h2>NEW CASEs.</h2>

How to play for an odd Trick.

If you are elder hand, and have the ace, king, and three fmall trumps, with four fmall cards of another fuit, three fmall cards of the third fuit, and one fmall card of the fourth fuit, queftion, how are you to play? You are to lead the fingle card, which, if won by the laft player, induces him to play trumps, or to play to your weak fuit, in which you and your partner gain the ten ace.

The like Cafe for an odd Trick when your Partner is to lead.

Suppofe he plays the ace of the fuit of which you have only one, and proceeds to play the king of the fame fuit, and your right hand adverfary trumps it, with the queen, knave or ten, you fhould not over trump him, but throw away the fmalleft card of your weakeft fuit, as this will leave your partner the laft player, and give him the ten ace in your weak fuit.

H 2   *The*

The like Case, supposing you want four or five Points, and are elder Hand.

Play a small trump, and if your partner has a better trump than the last player, and returns the lead, put in your king of trumps, and then play the suit of which you possess four cards.

A second Case.

A and B are partners against C and D; twelve trumps are played out, and seven cards only remain in each hand, of which A has the last trump, and likewise the ace, king, and four small cards of a suit, question, whether A should play the ace and king of that suit, or a small one? Ans. A should play a small card of that suit, as it is an equal bet his partner has a better card in that suit than the last player; and in this case, if four cards of the suit happen to be in either of the adversaries hands, by this manner of playing, he will be enabled to make five tricks in that suit. Should neither of the adversaries have more than three cards in that suit, it is an equal bet that he wins six tricks in it.

If A and B are partners against C and D; and eight trumps have been played out, and A has four trumps remaining, C having the best trump, and is to lead, should C play his trump or not? No; because as he would have three trumps in A's hand, if A's partner has any capital suit to make, by C's keeping the trump in his hand, he can prevent his making that suit.

A Case of Curiosity.

Supposing three hands of cards, containing three cards in each hand, let A name the trumps, and let B choose which hand he pleases. A having the choice of either the other two hands, will win two tricks.—Clubs are trumps: first hand, ace, king, and

and fix of hearts; fecond hand, queen and ten of
hearts, with ten of trumps; third hand, nine of
hearts, with two and three of trumps; the firft hand
wins of the fecond, the fecond wins of the third,
and the third wins of the firft.

Laws of the Game at Whist, as played at White's and Stapleton's Chocolate Houfes, &c.

Of dealing.

Firft. If a card is turned up in dealing, it is in
the option of the adverfe party to call a new deal;
but if either of them have been the caufe of turning
up fuch card, in that cafe the dealer has his option.

Second. If a card is faced in the deal, the dealer
muft deal again, unlefs it is the laft card.

Third. Every perfon ought to fee he has thirteen
cards dealt; therefore if any one fhould happen to
have only twelve, and does not difcover it till feveral
tricks are played, and the reft of the players have
their proper number, the deal ftands good; and alfo
the perfon who plays with twelve cards, is to be pu-
nifhed for each revoke, in cafe he has made any;
but if any of the other players happens to have fourteen cards, in that cafe the deal is loft.

Fourth. The dealer fhould leave in view upon the
table the turn-up card, till it is his turn to play; and
after he has mixed it with his other cards, no one is
entitled to demand what card was turned up, but
may afk what are trumps. Hence follows that the
dealer cannot name a wrong card, which otherwife
he might do.

Fifth. None of the players ought to take up or
look at their cards, while any perfon is dealing; and
if the dealer fhould happen to mifs deal, he fhall deal
again, unlefs it arifes from his partner's fault; and
if a card is turned up in dealing, no new deal fhall
be

he called, unless the partner has been the cause of
it.

Sixth. A deals, and instead of turning up the
trump, puts the trump card upon the rest of his cards,
with the face downwards ;—he is to lose the deal.

Of playing out of turn.

Seventh. If any person plays out of turn, it is in
the option of either of the adversaries to call the card
played, at any time in the deal, provided his playing
it does not cause him to revoke : or if either of the
adverse party is to lead, he may desire his partner to
name the suit he chooses to have him lead ; and when
a suit is thus named, his partner must play it if he
has it.

Eighth. A and B are partners against C and D ;
A plays the ten of a suit, the adversary C, plays the
knave of the same suit ; but before D plays, his partner leads a thirteenth, or some other card ; the penalty shall be in the option of A or B, to win the
trick, if he can.

Ninth. A and B are partners against C and D ; A
leads a club, his partner B, plays before the adversary C ; in this case D has a right to play before his
partner C, because B played out of his turn.

Tenth. If the ace, or any other card of a suit is
led, and it should happen that the last player plays out
of his turn, whether his partner has any of the suit led
or not, provided he is not forced to revoke, he is neither entitled to trump it or win the trick.

Of revoking.

Eleventh. If a revoke should be made, the adversaries may add three to their score, or take three
tricks from the revoking party, or take down three
from their score ; and the revoking party, provided
they are up, notwithstanding the penalty, must remain at nine ; the revoke taking place of any other
score of the game.

Twelfth.

Twelfth. If any perfon revokes, and, before the cards are turned, difcovers it, the adverfe party may call the higheft or loweft card of the fuit led, or have the option to call the card then played, at any period of the deal, when it does not caufe a revoke.

Thirteenth. No revoke to be claimed till the trick is turned and quitted, or the party who revoked, or his partner has played again.

Fourteenth. If any perfon claims a revoke, the adverfe party are not to mix their cards, upon forfeiture of the revoke.

Fifteenth. No revoke can be claimed after the cards are cut for a new deal.

Of calling Honours.

Sixteenth. If a perfon calls at any point of a game, except eight, either of the adverfe parties may call a new deal; and they are at liberty to confult each other, whether they will have a new deal.

Seventeenth. After the trump card is turned up, no perfon muft remind his partner to call, on penalty of lofing a point.

Eighteenth. When the trump card is turned up, no honours in the preceding deal can be fet up, unlefs they were before claimed.

Nineteenth. If any perfon calls at the point of eight, and his partner anfwers, and both the oppofite parties have thrown down their cards, and it appears that the other fide had not two by honours, in this cafe they may confult with each other whether they will ftand the deal or not.

Twentieth. If any perfon calls at eight after he has played, it fhall be in the option of his adverfaries to call a new deal.

Of feparating or fhewing the Cards.

Twenty-firft. If any perfon feparates a card from the reft, the adverfe party may call it, provided they name it, and prove the feparation; but in cafe they call

call a wrong card, they are liable to have the higheſt
or loweſt card called in any ſuit led during that deal.

Twenty-ſecond. If any perſon throws his cards
upon the table, with their faces upwards, ſuppoſing
that he has loſt the game, the adverſaries are entitled
to call any of the cards when they judge proper,
provided they do not make the player revoke, and he
is not to take up his cards again.

Twenty-third. If any perſon is ſure of winning
every trick in his hand, he may ſhew his cards upon
the table; but he is then liable to have all his cards
called.

Of omitting to play to a Trick.

Twenty-fourth. A and B are partners againſt C
and D; A leads a club, C plays the ace of clubs, and
D, partner to C, takes up the trick, without playing any card, A and the reſt of the players play on,
till it appears D has one card more than the reſt; the
penalty is in the option of the adverſaries to call a
new deal.

Reſpecting who played any particular Card.

Twenty-fifth. Each perſon in playing, ſhould lay
his card before him; after which, if either of the
adverſe party mixes his card with the player's, who
purſued this method, his partner is entitled to demand each perſon to lay his card before him, but not
to enquire who played any particular card.

*Rules to play any Hand of Cards, according to the neareſt
Calculations of your Partner's holding certain winning Cards.*

1. That he has not one certain winning
   card ·   —   —   2 to   1
2. Not two certain winning cards   17 to 2
   but it is about 5 to 4 that he has one
   or both, or   —   32 to 25
3. That he has one card out of any three
   certain winning cards, about ·   5 to   2
                                       4. That

4. That he has not three certain winning cards, about 31 to 1, or -   681 to 22

5. That he has not two of them,
about 7 to 2, or   —   —   547 to 156.

6. That he has not one of them,
about 7 to 6, or   —   —   378 to 325

7. That he holds one or two of them,
is in his favour about 13 to 6, or   481 to 222

8. And about 5 to 2 that he holds
one, two, or all three of them.

The odds of the game calculated with the deal.

The odds in favour of the deal at starting, are   —   —   21 to 20

1 love	—	—	— 11 to 10
2 love	—	—	— 5 to 4
3 love	—	—	— 3 to 2
4 love	—	—	— 7 to 4
5 love, an even bet of the lurch	—	2 to 11
6 love	—	—	— 5 to 2
7 love	—	—	— 7 to 2
8 love	—	—	— 5 to 1
9 love, not quite 5 to 1, but about —	9 to 2
2 to 1	—	—	— 9 to 8
3 to 1	—	—	— 9 to 7
4 to 1	—	—	— 9 to 6
5 to 1	—	—	— 9 to 5
6 to 1	—	—	— 9 to 4
7 to 1	—	—	— 3 to 2
8 to 1	—	—	— 9 to 2
9 to 1 about	—	—	— 4 to 1
3 to 2	—	—	— 8 to 7
4 to 2	—	—	— 4 to 3
5 to 2	—	—	— 8 to 5
6 to 2	—	—	— 2 to 1
7 to 2	—	—	— 8 to 3
8 to 2	—	—	— 4 to 1
9 to 2	—	—	— 7 to 2

I

4 to 3

4 to 3	—	—	—	7 to 6
5 to 3	—	—	—	7 to 5
6 to 3	—	—	—	7 to 4
7 to 3	—	—	—	7 to 3
8 to 3	—	—	—	7 to 2
9 to 3 about	—	—	—	3 to 1
5 to 4	—	—	—	6 to 5
6 to 4	—	—	—	6 to 4
7 to 4	—	—	—	2 to 1
8 to 4	—	—	—	3 to 1
9 to 4 about	—	—	—	5 to 2
6 to 5	—	—	—	5 to 4
7 to 5	—	—	—	5 to 3
8 to 5	—	—	—	5 to 2
9 to 5	—	—	—	2 to 1
7 to 6	—	—	—	4 to 3
8 to 6	—	—	—	2 to 1
9 to 6	—	—	—	7 to 4
8 to 7 above	—	—	—	3 to 2
9 to 7 about	—	—	—	12 to 8

9 to 8, or rather 8 to 9, the odds being
in favour of 8 about 3 and a half
per cut, according to the nicest
calculation.

*Odds calculated for betting throughout the whole rubber
with the deal.*

If the first game of a rubber is won, with 9 love of the second, on the same side, the odds of the rubber are nearly	13 to 1
Suppose the first game, and 8 love of the second is got, the odds are rather more than —— ——	13 to 1

When

When the first game is won, and 7 love		
of the second, near — —	8 to 2	
Ditto, and 6 love of the second, about	6 to 1	
Ditto, and 4 love of the second, about	5 to 1	
Ditto, and 3 love of the second, about	9 to 2	
Ditto, and 2 love of the second, about	4 to 1	
Ditto, and 1 love of the second, about	7 to 2	

Odds against the dealer throughout the rubber.

With the first game, and 9 love of the		
second, about —— ——	11 to 1	
Ditto, and 8 love of the second, rather		
more than —— ——	11 to 1	
First game, 7 love of the second ——	9 to 1	
Ditto, and 6 love of the second ——	7 to 1	
Ditto, and 5 love of the second ——	5 to 1	
Ditto, and 4 love of the second ——	4 to 1	
Ditto, and 2 love of the second ——	7 to 2	
Ditto, and 1 love of the second, near ——	13 to 6	

The terms or technical expressions used at Whist.

Finessing, is endeavouring to gain a trick, in case the player has the best and third best of a suit, by playing the third best, and running the risk of his adversary's having the second best, which being two to one in the player's favour, he may judiciously risk the event.

Forcing, is compelling a partner or adversary to trump, of which he has none.

Long trump, is having one or more trumps in hand, when all the rest are played.

Loose card, is one of no value, and therefore the properest to throw away

Points, Ten of these constitute the game ; honours as well as tricks telling towards the game.

Tierce, is a sequence of any three successive cards in the same suit.

I 2

Quart,

Quart, is a fequence of any four fucceffive cards; from quart major, which is a feqoence of ace, king, queen and knave, in any fuit, to cards of the loweft value.

Quint, is upon the fame principle a feqoence of five cards.

See-faw, is when a fuit, or two fuits, are trumped by both partners, and they keep playing thofe fuits to each other alternately.

Score, is the ftate and account of the game till it's conclufion.

Ten-ace, is having the firft and third beft cards, and being the laft player, by which fituation, the adverfary muft inevitably lofe the laft trick, let what card may be played of the fuit. Example, if the player, who has ace and queen of any fuit, and his adverfary leads that fuit, he muft win two tricks by having the beft and third beft of the fuit played, and being the laft player.

We fhall conclude our Treatife on Whift, with what is ftiled a Technical Memory, or an Affiftant to know what cards have been played, and which remains in hand.

Let the player place the trumps to the left of all the other fuits in his hand, his beft or ftrongeft fuit next, his fecond beft next, and his weakeft laft on the right hand.

If in the courfe of play, he finds he has the beft card remaining of any fuit, he fhould place it to the right of them, as it muft certainly win a trick after all the trumps are played.

When he finds he is poffeffed of the fecond beft card of any fuit, to remember, let him place it on the right hand of that card he has already to remember as the beft card remaining.

If

If he has the third beſt card of any ſuit, he ſhould place a ſmall card of that ſuit between the ſecond beſt card and his third beſt.

In order to remember his partner's firſt lead, he ſhould place a ſmall card of that ſuit led, entirely to the left of the trumps or trump, in caſe he has but one,

When he deals, let him put the trump turned up to the left of all his trumps; and as it is a kind of rule, he ſhould keep this trump as long as he is able, it will be more out of the way, and eaſier for him to recollect.

How to diſcover when, and in what ſuit, an adverſary revokes.

The player ſhould ſeparate four of his tricks from the remainder, remembering the firſt of theſe four tricks to ſtand for clubs, the ſecond for diamonds, the third for hearts, and the laſt for ſpades. In caſe he ſuſpects the revoke to have been made in ſpades, ſeparate the fourth trick a ſmall diſtance from the other three; if in hearts, ſeparate the third and fourth from the firſt and ſecond, and in like manner the reſt. From theſe tricks, the player will receive aſſiſtance, as it were, alphabetically; ſuppoſing the firſt trick to ſtand for the letter A, ſo clubs beginning with C, they ſhould be neareſt to the firſt letter of the alphabet; diamonds beginning with D ſhould ſtand next; hearts and ſpades then come in turn; by which means he may very eaſily recollect the ſuit, in which he thinks the revoke has been made. And by removing theſe towards the adverſary he ſuſpects of having revoked, he will probably remember in which trick the revoke took place.

The Game of Quadrille.

THIS is a very fashionable game, particularly amongst the ladies. It is played by four persons, with forty cards, the residue of a whole pack, after the four tens, nines, and eights, are discarded. They are dealt three by three, and one round four, to the right hand player. The trump is made by him or her, who plays, with or without calling, by naming spades, clubs, diamonds, or hearts, and the suit so named become trumps. If the person who names the trump should mistake, and say spades instead of clubs, or if he names two suits, the first named are trumps.

The cards placed according to their natural value.

Hearts and Diamonds.	Spades and Clubs.
king	king
queen	queen
knave	knave
ace	seven
deuce	six
four	five
five	four
six	three
seven	deuce
Total ten.	Total nine.

The reason that the aces of spades and clubs are not mentioned, is because they are invariably trumps, let whatever suit be played. The ace of spades being always the first, and the ace of clubs the third trump.

The

The cards placed according to their value when trumps.

Hearts and Diamonds.	Spades and Clubs.
Spadille, the ace of spades.	Spadille, the ace of spades.
Manille the seven of hearts or diamonds.	Manille, the two of spades or clubs.
Basto, the ace of clubs.	Basto, the ace of clubs.
Pont, the ace of hearts or diamonds.	
king	king
queen	queen
knave	knave
deuce	seven
three	six
four	five
five	four
six	three
Total twelve.	Total eleven.

Hence it is clear from the preceding tables, that spadille and basto are always trumps, and that the red suits have one trump more than the black.

There is a trump between spadille and basto, which is called manille, and is in black the deuce, in red the seven; they are the second cards when trumps, and the last in their respective suits when not trumps. Example: the deuce of spades being second trump, when they are trumps, and the lowest card when clubs, hearts, or diamonds are trumps, and so of the rest.

Ponto is the ace of hearts or diamonds, which are above the king, and the fourth trump, when either of those suits are trumps; but are below the knave, and called ace of diamonds or hearts, when they are not trumps. The two of hearts or diamonds is always superior to the three, the three to the four, the four to the five, and the five to the six; the six

is

is only fuperior to the feven when it is not trumps, for when the feven is manille, it is the fecond trump.

The three matadores, as they are called, are fpa-diile, manille, and bafto, whofe privilege is, when the player has no other trumps but them, and trumps are led he is not obliged to play them, but may play what card he thinks proper, provided, however, that the trump led is of an inferior value; but if fpadille fhould be led, he that has manille or bafto only, is compelled to play it, which is the cafe with bafto in refpect to manille, the fuperior matadore always for-cing the inferior. Although, properly fpeaking, there are but three matadores, yet all thofe trumps which fucceed the three firft without interruption, are alfo called matadores; but the three firft only en-joy the privilege above ftated. The number of the matadores are fpecified in the fecond table above, by the order and rank of the cards when they are trumps,

Of the manner of playing the game and dealing the cards, of the ftakes, of the manner of fpeaking, and of the beaft.

Every one is to play as he thinks proper, and moft advantageoufly to his own game.

He is not to encourage his friend to play; but each perfon fhould know what to do, when he is to play.

The ftakes confift of feven equal billets or con-tracts, as they are fometimes called, comprifing the ten counters and fifhes, which are diftributed to each player. A mille is equal to ten fifh, and every fifh to ten counters: the value of the fifh is according to the players agreement, as alfo the number of tours, which are ufually fixed at ten, and marked by turning the corners of a card.

Should the cards be wrong dealt, or fhould there be two of the fame fuit, as for example, two deuces

of

of fpades, there muft be a frefh deal, provided the
miftake is difcovered before the cards are all played.

A new deal muft likewife take place if a card is
turned in dealing, as it might be prejudicial to him
who might have it; and if there fhould be feveral
cards turned, the fame muft take place. No penalty
is inflicted for dealing wrong, but the dealer muft
deal again.

Each player having got his ten cards, he that is
on the right hand of the dealer, after examining his
game, and finding his hand proper to play, muft afk if
they play; or, if he has not a good hand, he paffes,
and fo the fecond, third, and fourth. All four may
pafs: but he who has fpadille, after having fhewn or
named it, is compelled to play, by calling a king.

If the deal is played in this manner, or one of the
players has afked leave, and no one choofing to play
without calling, the eldeft hand muft begin, previoufly
naming his fuit, and the king he calls: he who wins
the trick muft play another card, and the reft of
courfe till the game is finifhed. The tricks are then
reckoned, and if the *ombre*, meaning him who ftands
the game, has, together with him who has king called, fix tricks, they have won, and are accordingly
paid the game, the confolation and the matadores, if
they have them, and divide what is upon the game,
and the beafts, if any there be.

Should they make only five tricks, it is a *remife*, and
they are beafted, what goes upon the game, paying
to the other players the confolation and the matadores. When the tricks are equally divided between
them, they are alfo beafted; and if they make only
four tricks between them, it is a *remife*. Should
they make lefs, they lofe *codille*, and in that cafe pay
their adverfaries what they fhould have received if
they had won, namely, the game, confolation, and

matadores,

matadores, if they have them, and are beasted what is upon the game; and if they win codille, divide the stakes. The beast, and every thing that is paid, arise equally from the two losers; one half by him who calls, and the other moiety by him who is called; equally the same in case of *codille* as a *remise*, unless the *ombre* does not make three tricks, in which case, he who is called is not only exempt from paying half the beast, but also the game, consolation, and matadores, if there are any, which in that case the *ombre* pays alone, and likewise in case of a *codille* as a *remise*. This rule is enforced to prevent unreasonable games being played.

A singe case may occur, in which if the *ombre* makes only one trick, he is not beasted alone, which is when not having a good hand, he passes, and all the other players have passed likewise, and he having spadille is compelled to play. In this case it would be unjust to oblige him to make three or four tricks; wherefore, he who is called pays a moiety of the losing; and for the same reason, he who has spadille, with a bad hand, should pass, in order that if he is afterwards obliged to play by calling a king, (which is called forced spadille), he may not be beasted singly.

The player who has once passed, cannot be allowed to play; and he who has asked leave cannot refuse to play, unless another should propose playing without calling.

When a person has four kings, he may call a queen to one of his kings, but not that which is trumps. He who has one or more kings, may call one of those kings; but in this case, he must make six tricks alone, and therefore wins or loses singly. The king of the suit in which he plays cannot be called.

No one should play out of his turn, although he is not beasted for the trespass.

When

When he who is not eldeſt of hand, has the king called, and plays ſpadille, manille, or baſto, or even the king called, in order to ſhew that he is the friend, having other kings that he is apprehenſive the ombre may trump, he is not to be allowed to go for the vole ; and he is beaſted if it ſhould appear it is done with that deſign.

No hand is allowed to be ſhewn, though *codille* may already be won, in order that it may be ſeen whether the *ombre* is beaſted ſingly.

Should the ombre or his friend ſhew his cards, before he has made ſix tricks, judging that he might have made them, and there ſhould appear a poſſibility of preventing his making them, the other players may compel him to play his cards in what order they chooſe.

It is only neceſſary for a player to name his ſuit, when he plays, without calling a king.

Whoever plays without calling, muſt himſelf make ſix tricks to win ; all the other players being united againſt him, and therefore exert their combined efforts to diſtreſs him.

Whoever plays without calling, is permitted to play in preference to any other who would play with calling : neverthelefs, if he who has aſked leave, will play without calling, he has the preference of him who would force him. Theſe are the two methods of playing without calling, which are called *forced*.

He who plays without calling, not dividing the winnings with any other player, conſequently when he loſes pays all himſelf. Should he loſe by *remiſe*, he is beaſted, and pays each other player the *conſolation*, the *ſans appeller*, (commonly erroneouſly called the *ſans prendre*,) and the *matadores*, ſhould there be any. Should he loſe *codille*, he is alſo beaſted, and pays each player what he would have received from them if he had been the winner. Thoſe who win *codille* divide the gains; and if there are any remaining counters, they belong to the player of the three who may have spadille, or the highest trump in the succeeding deal. The same rule operates with respect to him who calls one of his own kings, he wins or loses alone, as in the other case, except the *sans appeller*, which he pays if a loser, or receives as a winner, although he plays singly.

Should he play *sans appeller*, though he may have a sure game, he is compelled to name his suit, which neglecting, shewing his cards, and saying *I play sans appeller*, in this case either of the rest of the players can oblige him to play in which suit he chooses, though he should not have a trump in that suit.

When a person has asked leave, he is not allowed to play *sans appeller*, unless he is *forced*: in this case, as beforementioned, he has the preference of the other players, by whom he is forced.

No player is compelled to trump, when he is not possessed of any of the suit led, nor obliged to play a higher card in that suit if he has it, it being optional to him, although he is the last player, and the trick belongs to the *ombre*: but he is compelled to play in the suit led if he can, otherwise he renounces.

Should he separate a card from his game and shew it, he is compelled to play it; if, by not doing it, the game should be prejudiced, or give any intelligence to his friend, but particularly if it should be a matadore. He who plays *sans appeller*, or by calling himself, is not subject to this rule.

One player may turn the tricks made by the others, and reckon what has been played, each time only it is his turn to play.

Should he, instead of turning a player's tricks, turn and see his game, or shew it to the other players, he is beasted, together with him whose cards he turned, each paying a moiety of the loss.

He

He who renounces, is beafted as often as detected; but no renounce takes place till the trick is turned.

Should the renounce be difcovered before the deal is finifhed, and has proved detrimental to the game, the cards muft be taken up again, and the game re-played from that trick where the renounce began. But fhould all the cards be played, the beaft ftill is made, and the cards muft not be re-played, unlefs there fhould be feveral renounces in the fame deal. In this cafe they are to be played again, un-lefs the cards fhould have been previoufly mixed to-gether.

When feveral beafts occur in the fame deal, they all go together, unlefs a different agreement is made; and in cafes of beafts, the greateft is firft reckoned.

Technical Dictionary of the phrafes ufed at Quadrille.

To afk leave, is playing by calling a king.

Beaft, is a penalty of paying as many counters as are down, incurred either by renouncing, or fome other fault; likewife by not winning when the player ftands his game.

Cheville, is being between the eldeft hand and the dealer.

Codille, is when thofe who defend the pool, make more tricks than thofe who defend the game, which is called winning the *codille*.

Confolation, is a claim in the game, always paid by thofe who lofe, whether by *codille* or *remife*.

Devole, is when he who ftands the game makes no trick.

Double, is to play for double ftakes, with regard to the game, *confolation* and *fans prendre*, *matadores* and *devole*.

Force, the ombre is faid to be forced, when a ftrong trump is played for the adverfary to over trump.

trump. He is likewife faid to be forced, when he
afks leave, and one of the other players obliges him
to play *fans prendre*, or pafs, by offering to play *fans
prendre*.

Friend, is the player who has the king called.

In paffe. To make the *in paffe*, is when being in
cheville, the knave of a fuit is played, of which the
player has the king.

Mille, is a mark of ivory, which is fometimes ufed,
and ftands for ten fifh.

Ombre, is the name given to him who ftands the
game, by calling or playing *fans appeller*.

Party, is the duration of the game, according to
the number of tours agreed to be played.

Pafs, is the term ufed when the players do not
choofe to play.

Pool, confifts of the fifhes, which are ftaked for
the deals, or the counters put down by the players,
or the beafts which go to the game. *To defend the
pool*, is to be againft him who ftands the game. *Pool*
likewife implies a certain number of counters, fupernumerary to the cards, when the tours are finifhed,
and the play is continued afterwards.

Prife, is the number of fifh or counters given to
each player at the commencement of the party.

Regle, is the order to be obferved at the game.

Remife, is when they who ftand the game, do not
make more tricks than they who defend the pool, and
they then lofe by *remife*.

Renounce, is not to follow the firft led, when at the
fame time, the player has a card of that fuit; likewife
when not having any of the fuit led, he wins with a
card that is the only one he has in the fuit which he
plays in.

Reprife, fynonimous to *party*.

Repert, the fame as *remife*.

Roi

Roi renda, is the king given up or surrendered ; in which case, the person to whom the king is given up, must win the game alone.

Forced spadille, is when he who has *spadille*, is obliged to play it, all the other players having passed.

Sans appeller, is playing without calling a king.

Sans prendre, is erroneously used for *sans appeller*, meaning the same.

Forced sans prendre, is when having asked leave, one of the players offers to play *sans prendre*, in which case he who asked leave, is obliged to play *sans prendre*, or pass.

Ten ace, is waiting with two trumps that must make, when he who has two others is obliged to lead.

Tours are the counters, which they who win by standing the game, put down to mark the number of *coups* played, by which the length of the party is determined.

The

The Game of Piquet.

PIQUET is played by two perfons only, with 32 cards, namely, the king, queen, knave, ten, nine, eight, and feven, of each fuit, which hold the fame rank as they are here ftated. In counting the game, the ace reckons eleven; the king, queen, and knave, ten each; and the fmall cards, according to the number of their pips.

After having fettled the fum to be played for, and the number of points that make the game, (which are ufually one hundred and one), the players cut for deal, and he who cuts the loweft piquet card deals firft, after having fhuffled the cards, and prefented them to his adverfary, who, if he pleafes, may alfo fhuffle them, in which cafe the dealer may fhuffle them again, when his antagonifts cuts. Should he cut only one, or drop a card in cutting, the dealer is authorifed to fhuffle once more.

The cards are dealt by two and two, and that way only. After twelve cards are dealt to each player, eight cards will remain, which are called the *talon*, heel or ftock, and remain on the table between the players.

If either of the players has thirteen cards dealt to him, it is at the option of the elder hand, either to play the cards, or call a new deal; and if he thinks proper to ftand the game, he is to lay out one card more than he takes in, in order that there may be three cards left the dealer. Should the younger hand have thirteen cards, he muft likewife lay out one card more than he takes in; but if either player has fourteen cards, there muft inevitably be a frefh deal.

To make the pique, the player muft be elder hand; for if he were the dealer, the elder hand would play a card and reckon one; and in that cafe, if the dealer

were

were to want 2**9** in hand, and win the card the elder
hand played, he would reckon only 30; unlefs the
elder hand play a card that did not reckon, as a nine,
eight, or feven; then the dealer, after having won the
trick, might go on to 30, reckon 60, and make the
pique.

The *carte blanche*, which is good for ten points,
counts firft, efpecially when the players are near the
conclufion of the game; after which follows the point
and fequence; and then the points which are told in
play; and laftly, the ten points for the cards, or 40
for the capot.

The *point* is the amount of fo many cards in one
fuit, the ace reckoning for eleven, the picture cards
for ten each, and the fmall cards according to the
number of their pips. When the point is reckoned
by the elder hand, he calls it, mentioning the amount,
and afking if it is good, to which his adverfary anfwers according to his hand; if he has not fo many,
he replies it is good; if he has an equal number, he
fays equal; and if he has a fuperior number, he fays
not good. The point is then reckoned by him who
has the fuperiority of number, containing as many
for the point as he has cards which conftitute the
whole; except, for example, he has fix cards that
reckon only forty-four, he reckons but five; whereas
had they made fifty-five, he would have counted fix;
and fo in refpect to fixty-four and forty-four, which
reckon no more than the number of their tens, as the
fifth point always maller up the ten, and thirty-five
points are equal in value to forty-four, each wanting
four. Neverthelefs in fome parties, they reckon every
card as one, whether the total is more or lefs than
forty-five, fifty-five, &c. Should the players points
be equal, no point is reckoned. If they hold the
fame fequence, the like rule is obferved, unlefs one of
the players fhould make his fequence good, by hold-

L

ing

ing a fuperior quint, quart, tierce, &c. than his adverfary.

General rules for playing the cards.

The primary object of a player fhould be to endeavour to make his fcore, that of twenty-feven points
elder hand, and thirteen points younger hand ; wherefore, if he has fix tricks, with a winning card in his
hand, he fhould invariably play that card, if he does
not difcover in the courfe of playing, what cards his
opponent laid out.

Should he be much advanced in the game, having, for example, attained eighty to fifty, it is judicious to let his adverfary gain two points for his one, as
frequently as he can, more particularly if he is to
become elder hand next deal ; but upon the fuppofi-
tion that he will be younger hand, and the game in
the fame fituation, he fhould not even in that cafe,
fear lofing two or three points to gain only one, as
that fingle point brings him within his fhew.

The elder, as well as younger hand, fhould fometimes fink a point, fuch as tierce, three kings, queens,
&c. in order to win the cards : but it fhould be done
judicioufly, and without hefitation, to fucceed.

It is alfo good play fometimes, for the younger
hand to fink one card of his point, which his adverfary may imagine to be a guard to a king or queen,
and thereby gain fuch an advantage in playing the
cards, as to obtain the majority of tricks.

To play judicioufly, the younger hand fhould in
moft cafes have his queen fuits guarded, with the
view of making points, and to fave the cards in playing them.

Should the elder hand be certain of making the
cards equal in playing them, and at the fame time
be more advanced in the game than his adverfary,
he fhould rifk the lofing of them ; but *vice verfa*, if
his

his adverſary ſhould be ſeveral points before him, he ſhould riſk the loſing of the cards, with the view of gaining them.

Laws of the game.

Firſt. If either of the players has thirteen cards dealt him, it is at the option of the elder hand, either to play the cards, or have a new deal, which ever he ſhall judge the moſt advantageous : but ſhould either of the players have fourteen cards, or upwards, a new deal muſt take place.

Second. Should the elder hand have thirteen cards, and chooſe to play them, he muſt put out one more than he takes in, as the younger hand muſt have his three cards, Should the younger hand have thirteen, the elder hand muſt take in the ſame cards as if the ſtock were right ; and the younger hand muſt lay out three, and take in two. When either caſe occurs, he who has thirteen cards, muſt inform his antagoniſt of his deſign before he takes in, as after he has ſeen his cards, the game muſt be played, under the penalty of playing twelve cards, which is reckoning nothing.

Third. The player who takes in more cards than he lays out, or in playing is found to have more cards than he is entitled to, reckons nothing ; whilſt his adverſary can count every thing he is poſſeſſed of, though inferior to what his opponent may have, under this circumſtance.

Fourth. Whoever plays with leſs than twelve cards, can reckon all he has, it being no fault to have too few cards ; but his adverſary always wants the laſt card, wherefore he cannot be capoted, when the other may for want of a twelfth card.

Fifth. The player who omits, at the beginning, to reckon *carte blanche*, his point on the ace, &c. or any fequence he may have good in his hand, cannot afterwards reckon them.

Sixth. Whoever forgets fhewing his point, fequence, &c. before he plays his firft card, which he may have better than his opponent, cannot count them afterwards. In this cafe the elder hand, whofe point, fequence, &c. or three of any fort, which were not allowed to be good, has a right, provided he has not played his fecond card, to reckon his game, which had not been ealled or fhewn.

Seventh. At the conclufion of each game, the players muft cut for deal, unlefs there is a previous agreement to deal alternately throughout the party.

Eighth. Neither players can difcard twice; and as foon as he has touched the ftock, whatever cards he has difcarded, cannot be again taken in.

Ninth. No player can fee the card he is to take in, before he has difcarded; wherefore, when the elder hand leaves any of the take-in cards, he muft fpecify what number he takes in, or how many he leaves.

Tenth. The player who has laid out lefs cards then he has taken in, and perceives his error before he has turned any of them, or mixed them with his own cards, is allowed to return the fupernumerary cards, without incurring any penalty, provided always that his adverfary has not taken in his cards, as in that cafe, he is at liberty either to play the cards, or to demand a new deal. Should the deal be played, the fupernumerary card muft be mixed with one of the two difcards, after being feen by the players.

Eleventh. Whoever deals twice fucceffively, and recollects himfelf before he has feen his cards, may compel his opponent to deal, though the latter has feen his cards.

Twelfth. Should the elder of hand call his point, or any thing elfe he may have to reckon, and his opponent reply it is good, but upon examination find

himfelf

himſelf miſtaken, he is allowed to reckon what he
has that is good, on condition that he has not played ; and likewiſe to ſet aſide what was called by the
elder hand, even though the firſt card ſhould have
been played.

Thirteenth. The player who might have quatorze
aces, kings, queens, knaves, or tens, and has diſcard-
ed one of them, and therefore reckons only three
aces, kings, &c. which are allowed to be good, muſt
tell his opponent with preciſion, what card he has
laid out, as ſoon as he has played his firſt card, provided he is aſked.

Fourteenth. Should the pack be erroneous, that is
to ſay, ſhould there be two tens, or any other two
cards of the ſame ſuit, or ſhould there be a card ſu-
pernumerary, or one deficient, the deal is void ; but
the preceding deal remains valid.

Fifteenth. If there ſhould be a faced card in dealing, there muſt be a freſh deal.

Sixteenth. If there ſhould be a faced card in the
ſtock, the deal muſt ſtand good, unleſs it is the upper card, or the firſt of the three that belong to the
dealer ; but in caſe of two faced cards, a new deal
neceſſarily enſues.

Seventeenth. He who calls his game wrong, and
does not correct himſelf before he begins to play,
reckons nothing he has in his game ; for if the adverſary diſcovers, at the beginning, middle, or end
of the deal, he ſhall not only prevent his adverſary
from reckoning, but he ſhall himſelf reckon all he has
good in his game, which the other cannot equal.

Eighteenth. Any card which is ſeparated, and has
touched the board, is deemed to be played. Never-
theleſs, if a card is played to the antagoniſt's lead,
of a ſuit different from what has been played, he is
entitled to take it up and play another of the proper
ſuit ; for as there is no penalty for a renounce,
there

there cannot be any for that; but if the player fhould have none of the fuit led, and plays a card he did not intend, he is not permitted to take it up again, after he has once quitted it.

Nineteenth. Whoever fays, *I play in fuch a fuit,* and afterwards does not play that fuit which he fhould play, in order to fee the cards the dealer has left, is liable to be compelled by his opponent to play in what fuit the latter choofes.

Twentieth. The player who, by accident, or otherwife, turns or fees a card appertaining to the ftock, is to play in what fuit his antagonift may fix upon.

Twenty-firft. The perfon who having left a card of the ftock, mixes it with his difcard, without fhewing it to his adverfary, is obliged, after having named the fuit he propofes leading, to fhew all his difcard.

The Game of Lanſquenet.

THE reader will plainly perceive that this is originally a French game. It may be played at by any indiſcriminate number of people, though a ſingle pack of cards is uſed during the deal. The dealer, who poſſeſſes an advantage, ſhuffles the cards, and after they have been cut by another of the party, deals out two cards on his left hand, turning them up, then one for himſelf, and a fourth that he places on the table for the company, which is called the *rejouiſſance*. On this card any, or all the company, the dealer excepted, may put their money, which the dealer is compelled to anſwer. The dealer continues turning the cards upwards, one by one, till two of a ſort come up, that is to ſay, two aces, two deuces, &c. which, to prevent miſtakes, or their being conſidered as ſingle cards, he places on each ſide of his own card ; and as often as two, three, or the fourth ſort of a card come up, he invariably places, as beforementioned, on each ſide of his own card. The company has a right to take and put money upon any ſingle card, unleſs the dealer's card ſhould happen to be double, which is often the caſe, by his card being the ſame as one of the two hand cards, which he firſt dealt out on his left hand : thus he continues dealing, till he brings either their cards or his own. Whilſt the dealer's own card remains undrawn, he wins ; and which ever card is turned up firſt, loſes. If he deals out the two cards on his left hand, which are ſtiled the hand cards, before his own, he is entitled to deal again. This advantage amounts to no more than his being exempted from loſing, when he turns up a ſimilar card to his own, immediately after he has turned up one for himſelf.

Lanſquenet

Lanfquenet is often played without the *rejouiffance*, the dealer giving every one of the party a card to put their money upon.   It is alfo often played by dealing two cards, one for the company, and the other for the dealer.

It fhould likewife be obferved, that a limitation is generally fixed for the fum to be placed upon any card or number of cards, either in gold or filver, beyond which the dealer is not obliged to anfwer.

The

The Game of Quinze.

QUINZE is another French game, and implies in English *fifteen*, which must be made as follows: First the cards must be shuffled by the two players, as that is the usual number who play at this game. After they have cut for the deal, which is determined by the lowest card, the dealer is authorised to shuffle them again ; after this the adversary cuts them, when the dealer gives one card to his opponent, and another to himself. Should the dealer's adversary not approve of his cards, he is entitled to have as many cards given him successively, as will make fifteen, or come nearest to that number, which are commonly given from the top of the pack. Example : if he should have a deuce, and draws a five, which amount to seven, he must continue going on, in expectation of coming nearer fifteen ; should he draw an eight, which make just fifteen, he, as eldest of hand, is certain of winning the game ; but should he over draw himself, and make more than fifteen, he loses, unless the dealer should do the same, which circumstance constitutes a draw game, and they consequently double the stakes ; thus persevering till one of them wins the game, by standing and being nearest fifteen. Upon the close of each game, the cards are packed and shuffled, when the players again cut for deal, the advantage being invariably for the elder hand.

M

The Game of E-O:

Entirely Original.

THIS very fashionable game, which is now played at most of the polite chocolate houses at the west end of the town, as well as Bath, Scarborough, Brighthelmstone, &c. has never yet been touched upon or explained, either by Hoyle, or his different supposed improvers. We, therefore, judged that some account of it here, would be far from proving disagreeable to our readers, as many, we imagine, may have no idea of the nature of the game, or the manner of playing it.

An E-O table is circular in form, of no exact dimensions, some tables being larger, others smaller, according to the size of the room it is played in, and the number of players that may be expected. In general, it is about a yard and one third in diameter. The extreme circumference is a kind of counter or *depot*, for the wagers or stakes, being marked all round with the letters E and O, on which each better puts his money according to his choice. The interior part of the table consists first of a kind of gallery or rolling place for the ball, which, with the outward parts, that we have distinguished by depot or counter, is stationary or fixed. The most interior part moves upon an axis, pivot, or spring, and is turned about with handles, whilst the ball is set in motion round the gallery. This part is divided generally into forty niches, or interstices, to receive the ball, twenty of which are marked E, and the other twenty O. The lodging of the ball in either of those letters determines the wager. Thus by there being two operations at once to determine the wager, (namely, first the circulation of the ball round the gallery, and its lodgment in one of these niches,

and

and the revolution of the interior table,) it should seem this must be the fairest game in the world, and that the player bets his money to no kind of disadvantage; but when it is recollected, the box, or proprietor, has a very extraordinary *pull*, this idea must vanish. Formerly this game was played on the same terms as Hazard, that is to say, who ever won, or threw in three times succeffively, paid, when gold was playing for, half a guinea to the proprietor of the table, or what was called the box. But the proprietors of the tables have now taken another method of paying themselves, by holding the box, and having two bar holes; according to which regulation, the boxholder is obliged to take all bets that are offered, either for E or O; but if the ball falls into either of the bar holes, he wins all the bets; which advantage is at the rate of 2 in 40, or 5 per cent. in his favour; a circumstance which in the long run, would exhaust the *Exchequer*.

Various collusions have also been detected; such as having a table constructed upon false principles, whereby the letter E or O, had all their niches larger than the other letter, and by that means eventually determined the bets in its favour. We have also heard of other artifices, such, as waxing a particular letter all round the table, and by that means stopping the progress of the ball, and fixing it to those particular niches.

We mention these artifices to put a player upon his guard; though, at the same time, we believe they are seldom practised.

THE END.

www.ingramcontent.com/pod-product-compliance
Lightning Source LLC
Chambersburg PA
CBHW021948190326
41519CB00009B/1190